AF361777

Immobilized Non-Precious Electrocatalysts for Advanced Energy Devices

Immobilized Non-Precious Electrocatalysts for Advanced Energy Devices

Editor

Nasser A. M. Barakat

MDPI • Basel • Beijing • Wuhan • Barcelona • Belgrade • Manchester • Tokyo • Cluj • Tianjin

Editor
Nasser A. M. Barakat
Organ Mat & Fiber Engn
Dept, Chonbuk Natl Univ,
Jeonju 561756, Korea

Editorial Office
MDPI
St. Alban-Anlage 66
4052 Basel, Switzerland

This is a reprint of articles from the Special Issue published online in the open access journal *Catalysts* (ISSN 2073-4344) (available at: https://www.mdpi.com/journal/catalysts/special_issues/ Immobilized_Electrocatalysts).

For citation purposes, cite each article independently as indicated on the article page online and as indicated below:

LastName, A.A.; LastName, B.B.; LastName, C.C. Article Title. *Journal Name* **Year**, *Volume Number*, Page Range.

ISBN 978-3-0365-4545-5 (Hbk)
ISBN 978-3-0365-4546-2 (PDF)

Contents

About the Editor

Nasser A. M. Barakat

Nasser A. M. Barakat (Professor) obtained his PhD in Chemical Engineering from Hunan University, China in 2004. From 2007 to 2009 he was a postdoctoral Research Fellow at the Bionanosyetm Engineering Department, Jeonbuk National University, Jeonju, South Korea. From 2010 to 2018, he was a faculty member in the Organic Materials and Fiber Engineering Department, Jeonbuk National University, Jeonju, South Korea. Currently, he is working as a professor in the Chemical Engineering Department, Minia University, Egypt. From 2013 to 2018, he was hired a visiting professor in King Saud University, Riyadh, Saudi Arabia. He was selected among the "World Ranking of Top 2% Scientists" in 2021, which was created by experts at Stanford University, USA; he has Rank # 374 (out of 4637) in Energy, and Rank # 683 (out of 6089) in Materials among the top 2% researchers list. According to the AD Scientific Index, in the field of chemical engineering, Prof. Barakat's rank is 1, 1, 2 and 359 in Minia university, Egypt, Africa and the world, respectively.

Preface to "Immobilized Non-Precious Electrocatalysts for Advanced Energy Devices"

Indeed, the precious metals possess fantastic catalytic and electrocatalytic activities; however, the rarity and consequent high cost of these metals forced the applicants to use nonprecious metals as alternatives in the research and industrial fields. Synthesis in nanoscale can distinctly enhance the catalytic activities of the nonprecious metals. The tiny size of the nanostructures makes using them in a free state a difficult task; therefore, the immobilization of the metallic nanostructures on proper supports became an urgent and hot research topic. Advanced energy devices are the most important target to utilize the immobilized nonprecious metals as electrodes. Consequently, writing a book that contains the up-to-date reported strategies in the immobilization of nonprecious metals' nanostructures is highly recommended. Based on the size and morphology of the nanostructures, the immobilization methodology and used support were decided. It was realized that carbonaceous supports, specifically graphene and carbon nanofibers, were widely exploited as supports due to their high adsorption capacity and excellent electric conductivity. Consequently, this book focused on the hot research dealing with these two carbonaceous supports. The editor has carefully selected pioneering researchers in the field of invoking the immobilized nonprecious metals as electrodes in advanced energy devices to contribute to this book. Overall, in this book, the readers will find a valuable collection of fantastic and up-to-date applications of immobilized nonprecious metals on graphene and carbon nanofiber supports as electrocatalysts.

Nasser A. M. Barakat
Editor

Editorial

Immobilized Non-Precious Electrocatalysts for Advanced Energy Devices

Nasser A. M. Barakat

Chemical Engineering Department, Minia University, Elminia 61519, Egypt; nasser1995@hotmail.com

Citation: Barakat, N.A.M. Immobilized Non-Precious Electrocatalysts for Advanced Energy Devices. *Catalysts* **2022**, *12*, 607. https://doi.org/10.3390/catal12060607

Received: 13 May 2022
Accepted: 25 May 2022
Published: 2 June 2022

Publisher's Note: MDPI stays neutral with regard to jurisdictional claims in published maps and institutional affiliations.

The expected near depletion of fossil fuels encourages both the research and industrial communities to focus their efforts to find effective and sustainable alternatives. Electrochemical devices, including fuel cells, batteries, supercapacitors, solar cells, etc., have been found to be the optimal life raft away from this dilemma. Aside from the potential for power generation from untraditional energy sources such as wastewater, electrochemical devices are highly recommended from an environmental point of view. The zero-emission of the exhaust gases greatly helps to solve one of the most dangerous problems currently facing the planet: the increasing climate temperature. Therefore, it would not be an exaggeration to claim that the development of electrochemical devices is considered an existential demand.

In the first era of these promised devices, precious metals such as platinum, palladium, and ruthenium were the main constituents in the manufacture of both cathodes and anodes. However, their high cost constrains the commercial application of these devices [1]. Compared to anodes, precious metal-based compounds as cathode materials were widely exploited. However, the thermodynamic potential of ORR (1.23 V vs. NHE at standard conditions) is so high that the Pt electrode cannot remain pure. Accordingly, the performance decreases due to the formation of PtO [2,3]. Consequently, non-precious metals must be used as electrodes in electrochemical devices not only for cost-decreasing reasons but also for the long-term use of these devices. In this regard, numerous non-precious electrodes have been introduced for different electrochemical devices including fuel cells [4–6], supercapacitors [7], batteries [8], and others [9–12].

Since most of the electrochemical reactions can be considered a combination of adsorption and chemical reaction, carbonaceous materials are widely invoked as supports for different functional materials. Specifically, carbon nanostructures including graphene, carbon nanotubes, and carbon nanofibers show a distinguished enhancement in electrocatalytic activity. Their large surface area and well-known high adsorption capacity improves both mass transfer and reaction(s) kinetics operations. Therefore, different techniques have been proposed for the immobilization of the functional materials on proper supports.

In this regard, this Special Issue entitled "Immobilized Non-Precious Electrocatalysts for Advanced Energy Devices" aims to collect innovative and high-quality research. Along with being a highly informative review article in the field of green hydrogen production, this issue contains nine articles introducing different functional non-precious materials immobilized on various supports. Moreover, the guest editor did his best to focus on popularly discussed topics to draw the maximum amount of attention from the researchers working in the advanced energy devices field. Consequently, a variety of interesting topics are covered in this issue including urea electrooxidation, dye sensitized solar cells, oxygen reduction reactions, and methanol & ethanol electrooxidation.

In summary, the current Special Issue covers cutting-edge techniques for addressing open problems about the use of non-precious metals in the most appealing energy devices and the immobilization of these functional materials on appropriate supports. Finally, the guest editor wishes to thank the editorial staff for their professional help as well as all the contributed authors for their outstanding scholarly contributions.

Conflicts of Interest: The author declares no conflict of interest.

References

1. Xu, H.; Ci, S.; Ding, Y.; Wang, G.; Wen, Z.J. Recent advances in precious metal-free bifunctional catalysts for electrochemical conversion systems. *J. Mater. Chem. A* **2019**, *7*, 8006–8029. [CrossRef]
2. Hoque, M.A.; Hassan, F.M.; Higgins, D.; Choi, J.Y.; Pritzker, M.; Knights, S.; Ye, S.; Chen, Z. Multigrain Platinum Nanowires Consisting of Oriented Nanoparticles Anchored on Sulfur-Doped Graphene as a Highly Active and Durable Oxygen Reduction Electrocatalyst. *Adv. Mater.* **2015**, *27*, 1229–1234. [CrossRef] [PubMed]
3. Hoare, J.P. *Electrochemistry of Oxygen*; Wiley: New York, NY, USA, 1968.
4. Kiani, M.; Tian, X.Q.; Zhang, W. Non-precious metal electrocatalysts design for oxygen reduction reaction in polymer electrolyte membrane fuel cells: Recent advances, challenges and future perspectives. *Coord. Chem. Rev.* **2021**, *441*, 213954. [CrossRef]
5. Ren, X.; Liu, B.; Liang, X.; Wang, Y.; Lv, Q.; Liu, A. Review-Current Progress of Non-Precious Metal for ORR Based Electrocatalysts Used for Fuel Cells. *J. Electrochem. Soc.* **2021**, *168*, 044521. [CrossRef]
6. Barakat, N.A.; El-Deen, A.G.; Ghouri, Z.K.; Al-Meer, S. Stable N-doped & FeNi-decorated graphene non-precious electrocatalyst for Oxygen Reduction Reaction in Acid Medium. *Sci. Rep.* **2018**, *8*, 3757. [PubMed]
7. Sawant, S.Y.; Han, T.H.; Ansari, S.A.; Shim, J.H.; Nguyen, A.T.N.; Shim, J.-J.; Cho, M.H. A metal-free and non-precious multifunctional 3D carbon foam for high-energy density supercapacitors and enhanced power generation in microbial fuel cells. *J. Ind. Eng. Chem.* **2018**, *60*, 431–440. [CrossRef]
8. Cao, R.; Lee, J.S.; Liu, M.; Cho, J. Recent Progress in Non-Precious Catalysts for Metal-Air Batteries. *Adv. Energy Mater.* **2012**, *2*, 816–829. [CrossRef]
9. Jiang, W.-J.; Tang, T.; Zhang, Y.; Hu, J.-S. Synergistic Modulation of Non-Precious-Metal Electrocatalysts for Advanced Water Splitting. *Acc. Chem. Res.* **2020**, *53*, 1111–1123. [CrossRef] [PubMed]
10. Chen, Y.; Zheng, Y.; Yue, X.; Huang, S. Hydrogen evolution reaction in full pH range on nickel doped tungsten carbide nanocubes as efficient and durable non-precious metal electrocatalysts. *Int. J. Hydrogen Energ.* **2020**, *45*, 8695–8702. [CrossRef]
11. Barakat, N.A.M.; Abdelkareem, M.A.; Abdelghani, E.A.M. Influence of Sn Content, Nanostructural Morphology, and Synthesis Temperature on the Electrochemical Active Area of Ni-Sn/C Nanocomposite: Verification of Methanol and Urea Electrooxidation. *Catalysts* **2019**, *9*, 330. [CrossRef]
12. El-Eskandarany, M.S.; Banyan, M.; Al-Ajmi, F. Synergistic Effect of New $ZrNi_5/Nb_2O_5$ Catalytic Agent on Storage Behavior of Nanocrystalline MgH_2 Powders. *Catalysts* **2019**, *9*, 306. [CrossRef]

Review

Review of Anodic Catalysts for SO$_2$ Depolarized Electrolysis for "Green Hydrogen" Production

Sergio Díaz-Abad, María Millán, Manuel A. Rodrigo and Justo Lobato *

Enrique Costa Novella Building, Av. Camilo Jose Cela n 12, Chemical Engineering Department, University of Castilla-La Mancha, 13004 Ciudad Real, Spain; sergio.diazabad@uclm.es (S.D.-A.); maria.millan@uclm.es (M.M.); manuel.rodrigo@uclm.es (M.A.R.)
* Correspondence: justo.lobato@uclm.es; Tel.: +34-926-295-300

Received: 29 November 2018; Accepted: 21 December 2018; Published: 9 January 2019

Abstract: In the near future, primary energy from fossil fuels should be gradually replaced with renewable and clean energy sources. To succeed in this goal, hydrogen has proven to be a very suitable energy carrier, because it can be easily produced by water electrolysis using renewable energy sources. After storage, it can be fed to a fuel cell, again producing electricity. There are many ways to improve the efficiency of this process, some of them based on the combination of the electrolytic process with other non-electrochemical processes. One of the most promising is the thermochemical hybrid sulphur cycle (also known as Westinghouse cycle). This cycle combines a thermochemical step (H$_2$SO$_4$ decomposition) with an electrochemical one, where the hydrogen is produced from the oxidation of SO$_2$ and H$_2$O (SO$_2$ depolarization electrolysis, carried out at a considerably lower cell voltage compared to conventional electrolysis). This review summarizes the different catalysts that have been tested for the oxidation of SO$_2$ in the anode of the electrolysis cell. Their advantages and disadvantages, the effect of platinum (Pt) loading, and new tendencies in their use are presented. This is expected to shed light on future development of new catalysts for this interesting process.

Keywords: Green Hydrogen; SO$_2$ electrolysis; electrocatalysts; Westinghouse cycle

---

1. Introduction

Nowadays, hydrogen is considered an actor with a significant role in tackling climate change and poor air quality. Hydrogen can, in particular, be produced from a broad range of renewable energy sources, acting as a unique energy core providing low or zero emission energy to all energy consuming sectors. Thus, Europe, through the Fuel Cells and Hydrogen Joint and Undertaking (FCH-JU) assessed different "Green Hydrogen" pathways, which as of today have already become nearly commercial, have already undergone extensive research in past or ongoing programs in Europe, or are internationally providing evidence of their general promise for be ready for commercialization. Figure 1 shows the different processes for producing hydrogen from renewable energy sources, and the ones which were selected as more promising after a deeper evaluation [1].

On the way to a "hydrogen economy" [2] and "Green hydrogen", the ideal raw material is, obviously, water. However, the single step thermal dissociation of water to hydrogen and oxygen is one of the most challenging processes for producing hydrogen in practice [3], due to its unfavorable thermodynamics.

There are different routes to produce hydrogen with low or zero CO$_2$ emissions, including promising processes based on water splitting (in which new catalysts are being developed)—photoelectrochemical water splitting [4–6], thermochemical cycles [3,7,8] or hybrid thermochemical cycles [9,10] are some examples. Hybrid thermochemical cycles using a high temperature thermal source have been proposed as one of the most promising technologies for massive hydrogen production [11]; moreover, they have been included as one of the candidates for "Green hydrogen" production, as it can be seen in Figure 1.

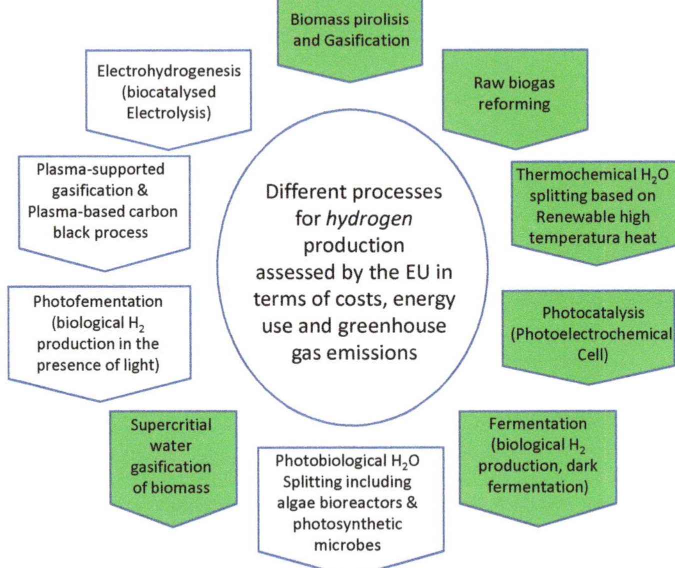

Figure 1. Proposed and selected (highlighted in green) "Green Hydrogen" production pathways in the European Union (EU) (adapted from [1]).

One of the leading thermochemical cycles to produce hydrogen with a high sustainability is the hybrid Sulphur cycle, also known as the Westinghouse cycle. It is a hybrid electrochemical–thermochemical cycle [9]. It was originally proposed in 1975 [12] and developed by Westinghouse electric corporation. The process is labelled "hybrid" because of the substitution of one thermochemical reaction by the electrochemical oxidation of SO_2 with water to yield sulphuric acid and hydrogen [3,13]. What makes this process interesting is its low theoretical cell potential of 0.158 V (versus SHE), which when compared with direct water electrolysis (E^0 = 1.23 V versus SHE) makes the Westinghouse process a great alternative in terms of electric energy consumption [3].

Figure 2 shows a possible configuration of the Westinghouse cycle. It consists of two main steps. In the first step, SO_2 is electrochemically oxidized at the anode to form sulphuric acid, protons and electrons (E^0 = 0.158 V versus SHE). The protons are conducted across the electrolyte separator to the cathode, where they recombine with the electrons to form hydrogen, according to Equations (1) and (2) [3,13–16].

$$SO_2(aq) + 2H_2O \rightarrow H_2SO_4(aq) + 2H^+ + 2e^- \tag{1}$$

$$2H^+ + 2e^- \rightarrow H_2(g) \tag{2}$$

The second step, common to all sulphur-based thermochemical cycles, is the result of two successive reactions. As sulphuric acid is vaporized (ca. 650 K) and superheated (ca. 900 K), it decomposes into water and sulphur trioxide. The following reaction is the catalytic decomposition, at temperatures higher than 1000 K, of the SO_3 to produce oxygen and sulphur dioxide (Equations (3) and (4)) and downstream are separated [3,13–15].

$$H_2SO_4(aq) \rightarrow SO_3 + H_2O(g) \tag{3}$$

$$SO_3 \rightarrow SO_2 + 1/2O_2 \tag{4}$$

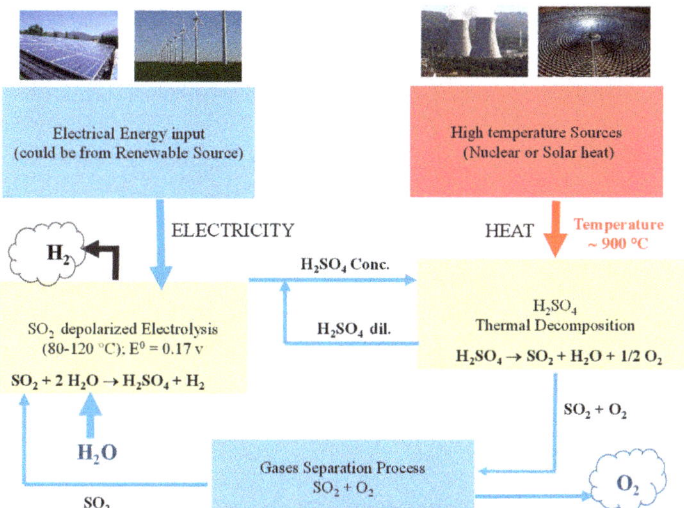

Figure 2. Scheme of the Westinghouse cycle.

The sulphuric acid decomposition has been widely investigated in different studies, such as the one from Sandia National Laboratories [11], and most efforts have been devoted to the optimization of the SO$_2$ depolarized electrolysis (the electrochemical stage) to increase the global efficiency of the cycle [16]. Challenges for this process are the reduction of the overpotentials in the electrolysis step in order to improve the overall efficiency of the process—the development of high corrosion resistance materials due to the use of acid solutions. Also, as for the thermochemical step in which H$_2$SO$_4$ has to be decomposed at very high temperatures, only high-temperature energy sources suitable for this process, such as concentration solar energy or nuclear energy. In terms of efficiency and costs, an optimal efficiency is 47% for a nuclear heat source with a hydrogen cost of 4 \$/kg [3].

The first design of the SO$_2$ electrolyzer was developed by Westinghouse Corp. in the 1970s, and it consisted of a conventional electrolysis cell with two compartments separated by a membrane [17]. Modern electrolyzers for this process are based on proton exchange membrane fuel cell (PEMFC) technology, where the membrane–electrode–assembly (MEA) is the core of the cell [18]. Figure 3 shows a scheme of a PEM electrolyzer for SO$_2$ depolarized electrolysis.

As mentioned before, the electrolysis step is one of the parts that has to be optimized. One of the proposed goals for this step is to operate at 0.6 V and 0.5 A/cm^2 [19]; however, this has not been yet reached.

The anodic electrocatalysts are the key factor to improve SO$_2$ electrolysis efficiency, as it is necessary to find catalysts with not only high activity but high stability and low costs. At the beginning, anodized electrodes with noble metals like Pt, Pd, Rh, Au, Ru, Re, or Ir were investigated, being the first to achieve better results [20,21]. Now the catalyst technology used in PEMFC systems has been copied, and the electrodes are based on a catalyst supported by carbon-based materials deposited on gas diffusion layers.

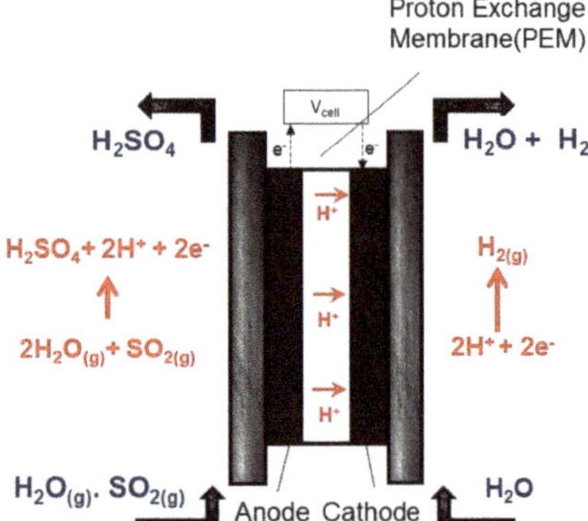

Figure 3. Scheme of a SO_2 depolarized electrolysis cell based on proton exchange membrane (PEM) technology.

This manuscript aims to provide a review on recent developments in electrocatalysts for SO_2 depolarization electrolysis. To the authors' knowledge, there is no review on this topic. In addition, taking into account the great impulse that renewable energies have in the world, and that the hydrogen produced with zero emission technologies can be one of the actors in the future energy scenario, this work can give an overview of the current status and future prospects of the catalysts for improving the electrochemical stage of the hybrid sulphur cycle.

2. Platinum-Based Catalysts

Platinum, as noble metal, is a well-known catalyst that has been widely used in fuel cells. It was one of the first studied catalysts for SO_2 electrolysis to produce hydrogen, showing great catalytic performance for the electro-oxidation of SO_2; however, its high price is a disadvantage for this material. For this reason, in recent years different materials have been examined and compared with this noble metal, in order to find a better option to be employed as catalyst.

Seo and Sawyer mention the necessity of pretreating the electrode to activate it, forming an oxide layer at high potentials, which is then stripped when negative potentials are applied to obtain the desired surface characteristics [22]. This paper also shows that the initial preconditioning of the electrode affects its performance. They recommend choosing an initial voltage of -0.15 V (versus Saturated Calomel Electrode, SCE). In a later study [23], the authors investigated the oxidation mechanism of SO_2 on platinum electrodes, demonstrating that the process is mass-diffusion controlled and is an electrochemical–electrontrasfer process at potentials lower than those needed for platinum oxide formation (<0.42 V versus SCE). Whereas, a chemical reaction occurs between sulfite and anodically-formed oxide at higher potentials. The same result regarding the oxidation mechanism on platinized platinum supported on porous carbon electrodes was obtained by Wiesener [24]. Appleby and Pinchon compared the catalytic activity of pure black platinum with other noble metals supported on carbon materials, using a rotating-disk paste electrode at 3000 rpm [21]. They obtained 0.3 A/cm^2 with 10% Pt on Norit BRX at 0.7 V (versus SHE), which they considered a good result. Nevertheless, pure black platinum showed a better catalytic activity, with approximately 0.5 A/cm^2 at 0.7 V (versus SHE). Platinum supported on 2 μm graphite spheres showed poor activity, probably due to lower surface area, demonstrating that the SO_2 oxidation is highly substrate-dependent. In this work, they suggest using an acid concentration of 50 wt % for optimal results. Lu and Ammon used 50 wt % H_2SO_4 at 25 °C and atmospheric pressure for SO_2 electro-oxidation [20]. Pre-anodized electrodes showed reproducible data when being pre-treated, as mentioned before [22]. The authors obtained a limiting current density of only 1.2 mA/cm^2 at 0.8 V (versus SHE) for a platinum electrode. In a different study, Lu and Ammon compared the performance of a Pt-catalyzed carbon plate electrode with a carbon cloth-supported electrode in an SO_2-depolarized electrolyzer [25]. They developed an electrolyzer with a loading of 7 mg/cm^2 of platinum on the anode and with a potential of 0.77 V (versus SHE) at a current density of 0.2 A/cm^2 in a new electrolyzer configuration (50 wt % H_2SO_4; 50 °C, atmospheric pressure). In a carried-out endurance test for 80 h at 0.1 A/cm^2, the stabilized voltage was ca. 675 mV, obtaining a hydrogen purity of 98.7%.

Scott and Taama used a Pt/Ti electrode for the electrolysis of SO_2 for testing its current efficiency [26]. At a current density of 10 mA/cm^2 and 20 °C, the current efficiency was initially greater than 95% and it decreased with time as the concentration of sulphite fell, indicating that the oxidation of sulphite is almost entirely controlled by mass transfer. Recent works have generally involved high active surface area electrodes, where the catalyst is loaded on dispersed conductive carbon particles. Thus, Weidener et al. [27] prepared a membrane electrode assembly (MEA), spraying an ink containing 40 wt % Pt onto the gas diffusion layer until desired loading was achieved. They obtained a voltage of 0.7 V at 0.3 A/cm^2 for a catalyst loading of 1 mgPt/cm^2 and 0.83 V at 0.3 A/cm^2, using the half Pt loading. Steimke and Steeper compared the performance of two different cells [28]. One of them had porous titanium as the electrode in both the anode and cathode, with platinum black as catalyst on the titanium surface, a loading of 4 mgPt/cm^2, and a specific area for the catalyst of 25 m^2/g. In the second cell, platinum was added to the Nafion membrane. The authors prepared a slurry consisting of 40 wt % platinum on carbon and Nafion solution, which was hot pressed onto the Nafion membrane used in the MEA configuration, the platinum loading for the anode and the cathode was 0.5 mgPt/cm^2. For the second cell, the authors obtained the lowest cell voltages for the highest tested sulfur dioxide concentration and the highest anolyte flowrate. Later, the same authors [29] tested six different MEA configurations for the electrolysis of SO_2. The best results were obtained for a Pt loading of 0.88 mgPt/cm^2 onto the anode side with 0.75 mV at 0.3 mA/cm^2 at 4 bar and 70 °C. For this study, the catalyst containing ink was sprayed on the shiny Teflon-coated side of each gas diffusion layer, until the required Pt loading was achieved. Colón-Mercado and Hobbs compared the catalyst activity of platinum supported on carbon and a pure platinum black electrode [30]. When the Pt was to be supported on carbon, an ink was prepared and placed onto the gas diffusion layer. They also studied the electrocatalytic activity of platinum at different temperatures and acid concentrations. For example, potentials for Pt/C were measured at 0.51 V, 0.56 V and 0.63 V versus SHE in 3.5 M, 6.5 M and 10.4 M H_2SO_4 solutions, respectively, and it exhibited instability in very high H_2SO_4 concentrations (10.4 M) at temperatures of 50 °C and above.

Allen et al. [31] observed that the response of a platinum disk electrode was different depending on the lowest minimum potentials applied in a cyclic voltammetry. Three main oxidative peaks were obtained. The first peak appeared before the formation of platinum oxide which occurs at potentials above 0.90 V (SHE), meaning that peaks II and III were influenced by the formation of these compounds. When the initial voltage was lower than 0.2 V the oxidation scenario was defined to some extent by acid concentration. Furthermore, by increasing acid concentration, the reaction was inhibited.

2.1. Effect of Platinum Loading

At the beginning, the Pt loading in the tests performed by Westinghouse in the 1980s was as high as 10 g Pt/cm^2. However, new electrode configurations allowed them to reduce the catalyst loading to 1 mgPt/cm^2, with no penalty on cell performance [17].

Nowadays, as in the case of the PEMFC technology, the aim is to use the minimum Pt loading possible, in order to minimize the costs of the system. As a consequence, the effect of Pt loading has been studied by different research groups. Thus, the effect of Pt loading as a catalyst was studied by Lee et al. [32]. They used a three-electrode electrochemical cell with H_2SO_4 as the electrolyte. Five different Pt loadings were tested (0.40, 0.80, 1.30, 2.34 and 4.02 mg/cm^2) by cyclic voltammetry (CV) in a deaerated 4.8 wt % H_2SO_4 solution. The current level of the CVs increased with an increase in the Pt loading amount over the whole potential range. Figure 4 shows the evolution of the current density for a value of voltage depending on the Pt loading.

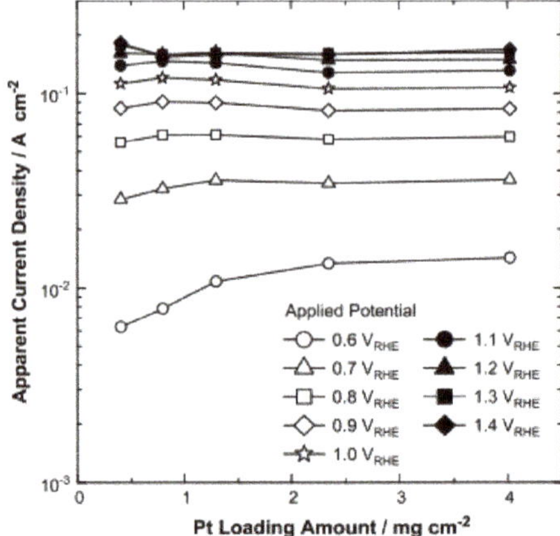

Figure 4. Apparent current density in the potential range of 0.6–1.4 V against platinum (Pt) loading [32]. Reprinted by permission of [24]. Copyright Elsevier Science BV. 2009.

The catalyst electrochemical active surface area (ESA) was calculated with these data, and it was concluded that the ESA did not increase at the same level as the Pt loading increased. This means that utilization of Pt decreased with loading, due to higher inactive sites (interfaces between particles, support and binding material). A study of the effect of Pt loading on SO_2 oxidation was carried out in an SO_2-saturated 50 wt % H_2SO_4 solution. The limiting current density increased as the Pt loading increased. For a cell potential of 0.6 V, the apparent current density increased from 6.3×10^{-3} to 1.4×10^{-2} A/cm^2 with an increasing Pt loading amount from 0.4 to 4.0 mg/cm^2. However, the current density remained almost constant at a high potential for the five studied catalyst loadings, indicating that the SO_2 oxidation reaction is controlled by the diffusion of dissolved SO_2 to the electrode in the high anodic over-potential range. Xue et al. [33] also carried out experiments with different Pt loadings in the range of 0.2–2.2 mg/cm^2. The higher loading amount led to better electrolysis performances, as a result of more electrochemical reaction sites, peaking at 1 mg/cm^2. Increasing the amount of platinum above that value led to worse performance, due to stacking of catalyst particles on one another. In addition, an excess of platinum increased the diffusion channel length and impeded the accessibility of the electro-active ion to the Pt surface [32,33]. For the authors' set-up, the optimal Pt loading was 1 mg/cm^2 in terms of platinum utilization, and polarization curves result. These authors also studied the influence of sulphuric acid concentration and temperature on the cell operating voltage, being the best electrolysis performance at 80 and 30 wt % sulphuric acid. Krüger et al. [34] evaluated some MEA manufacturing parameters, one of them being the catalyst loading. Their results are in concordance with studies carried out by previous authors [32,33]. Polarization curves obtained at 80 °C for catalyst loadings of 0.3, 0.5 and 1 mgPt/cm^2 showed that the best performance was for the catalyst loading of 1 mgPt/cm^2. However, it should be noted that for the catalyst loadings of 0.3 and 0.5 mgPt/cm^2, the platinum was supported on carbon, and for the highest catalyst loading platinum black was used.

Staser et al. [35] evaluated the effect of catalyst loading on the anodic overpotential suing a model and in the range of 0.001 to 1.5 mg Pt/cm^2. It was found that the anodic overpotential was mainly dependent on the slow oxidation kinetics, with ohmic losses and concentration losses comprising only a negligible fraction of the total losses. Moreover, at catalysts loadings below 0.1 mg/cm^2 the anodic overpotentials increased exponentially, and the optimum loading was found to be 0.2 mg/cm^2. However, no experimental tests have been found to support these results.

2.2. Combination of Platinum with Other Metals

As platinum has a high cost, the catalytic activity of some catalysts based on platinum but mixed with other metals have been investigated. The goal of this approach is to obtain catalysts as good as platinum, or even better materials for the electrochemical reduction of SO_2 with a lower cost, which would make it easier to scale the process.

Xue et al. [18] studied the electrochemical catalytic activity of different bi-metallic materials based on platinum. Vulcan XC-72R was the support for all catalysts. The metals that were employed were palladium, rhodium, ruthenium, iridium, and chromium. The total metal content was 60 wt %, with an atomic ratio of Pt–M 1:1. The catalyst loading was 1 mg metal/cm^2. Figure 5 shows the performance of each catalyst, and for comparison purposes, platinum supported on a Vulcan XC-72R is also shown. The catalysts were evaluated at room temperature and in 30 wt % sulphuric acid concentration. The results showed that the best catalyst was 60 wt % Pt-Cr/XC72R, which was better than the catalyst containing only platinum.

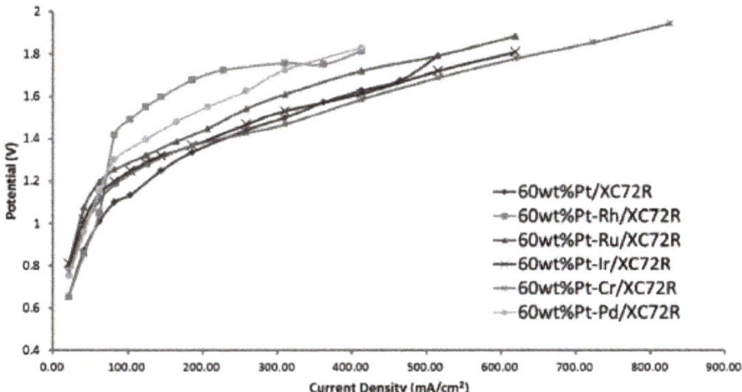

Figure 5. Polarization curves of Pt/C and Pt-based bimetallic catalysts for SO_2 depolarized electrolysis [18]. Reprinted by permission of [24]. Copyright Elsevier Science BV. 2014.

In this work [11], the influence of the atomic ratio was also studied. The ratio 1:2 (Pt:Cr) resulted in equal or even better electrolysis performance than that of 60 wt % Pt/C. Also, the catalyst Pt–Ir/C showed promising results.

Falch et al. [36] studied the possibility of using plasma sputtering to form a electrocatalytic film of platinum and palladium. Both metals were sputtered at the same time to obtain bimetallic materials. Catalysts with different molar ratios of Pt and Pd were prepared. The compositions ranged from a Pt composition of 0 (pure palladium) to 1 (pure platinum). This technique allowed homogeneous distribution of both metals in the surface. The onset potential of the prepared catalysts was measured, as it is an indicator of catalytic activity [37–39]. The combinations with the lowest onset potential were Pt_3Pd_2 (0.587 V), Pt_2Pd_3 (0.590 V) and $PtPd_4$ (0.587 V), which exhibited an onset potential slightly lower than that of pure platinum (0.598 V). The combination with the most promising results was Pt_3Pd_2, which was further investigated by the same authors [40]. The effect of thermal annealing on that sputtered catalyst and on pure platinum was investigated. Catalyst films were deposited as mentioned previously. The samples were annealed at temperatures ranging from 600 °C to 900 °C. The surface of the materials after deposition were smooth with no cracks; however, when increasing the temperature up to 900 °C, a discontinuous grain surface was formed. They measured the electrochemical surface area for pure platinum and the mixture Pt_3Pd_2 when it was rapidly annealed, and for a non-annealed sample. The annealing process clearly decreases the electrochemical active area, but for the Pt being smaller than for Pt_3Pd_2, this difference increased with an increase in the annealing temperature. Regarding onset potential, results showed that annealing does not have a positive influence on the catalyst, because it is higher than when they are not annealed for both platinum and Pt_3Pd_2. However, this technique does increase the lifetime of the catalysts in an acidic environment for Pt and Pt_3Pd_2 and normalised current density. The next step in their research was to add a non-noble metal, aiming better catalyst activity and a lower price. Falch et al. [41] developed a film composed of platinum, palladium and aluminium, which are co-sputtered on a support. They made improvements on the onset potential and in current density, obtaining 396.73 mA/mgPt with the ternary combination $Pt_{40}Pd_{57}Al_3$. Results changing Pt content show how the normalised current density increases with decreasing platinum content, from 100 mA/mgPt for pure platinum to 400 mA/mgPt for $Pt_{40}Pd_{57}Al_3$. They concluded that the addition of other metal enhanced the electrocatalytic performance. Also, when annealed the ternary catalyst $Pt_{40}Pd_{57}Al_3$ increases the amount of aluminium on the surface. Xu et al. [42] synthetized a $Pt/CeO_2/C$ catalyst for enhancing the SO_2 electro-oxidation. ESA measurements showed that adding CeO_2 increased the active area of the catalysts, i.e., $Pt/10CeO_2/C$ (810.60 cm^2/mg Pt) and $Pt/20CeO_2/C$ (765.10 cm^2/mg Pt), which almost doubled Pt/C (429.10 cm^2/mg Pt). The ESA of $Pt/30CeO_2/C$ was 512.90 cm^2/mg Pt and that of $Pt/40CeO_2/C$ was 428.40 cm^2/mg Pt. Catalysts with a content of CeO_2 below 50% gave current densities lower than for Pt/C, with the best ratio being $Pt/20CeO_2/C$. The enhanced SO_2 electrooxidation of $Pt/CeO_2/C$ composite catalysts was attributed to the oxygen provided by CeO_2, but no tests in an electrolysis cell were performed to support their findings of the half cell.

As summary, Figure 6 shows the values of voltage reached at different current densities obtained by different authors from 1980 until 2016, where it can be seen that there has been an improvement. However, in all cases the voltages are higher than the target proposed of 0.5 V at 0.5 A/cm^2.

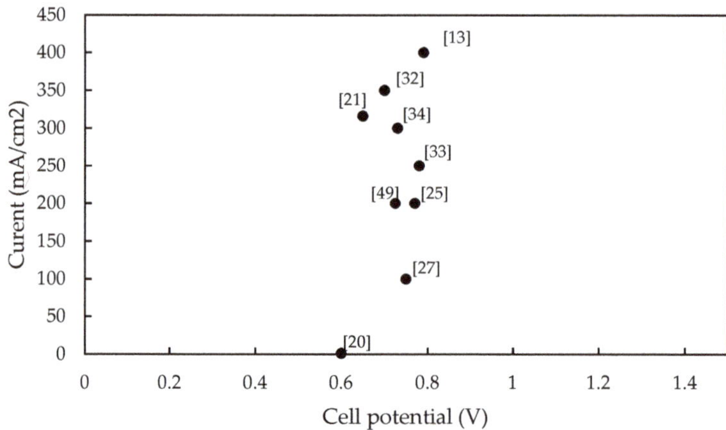

Figure 6. Evolution of cell performance for SO_2 electro-oxidation with platinum-based catalysts. Data taken from bibliography.

3. Gold-Based Catalysts

In the early development of the electrolyzer for the hybrid sulphur cycle, gold was studied together with platinum as a catalyst for SO_2 depolarized electrolysis.

Seo et al. [23] studied the electrochemical oxidation of sulphur dioxide on platinum and gold electrodes. The results on gold electrodes showed that reproducible data is obtained by scanning three or four times between −0.15 and 1.5 V (SCE). Compared with a platinum electrode, gold was easier to activate, due to the ease with which gold surface oxide films dissolve in acid solutions. However, extensive cathodization inhibited the electrode activity. They also distinguished two modes for the electrochemical oxidation of SO_2 on gold, a pure electrochemical process and a chemical oxidation process. Appleby et al. [21] determined that gold is less promising than platinum, because the oxidation of SO_2 involves participation of chemisorbed H, OH or O species. Rates for these processes are strongly influenced by the chemisorption properties of the substrate, and are several orders of magnitude higher on platinum than on gold. On the contrary, Samec and Weber [43] studied the oxidation of SO_2 dissolved in 0.5 M H_2SO_4 on a rotating disc gold electrode. These authors reported a considerable enhancement of the SO_2 oxidation reaction after a preliminary voltammetry cycle to potentials encompassing SO_2 reduction. They proposed that oxidation of SO_2 on gold electrodes proceeds through an adsorbed intermediate, which is displaced from the electrode at voltages higher than 1.5 V when the oxide is formed; the oxidation current decreases to a fraction of the limiting current density at that potential. The authors also mention that SO_2 oxidation is a mass transfer-controlled process to the electrode. Similar results to Seo et al. [23] were obtained by Lu and Ammon, which showed similar catalytic activity for gold and platinum with similar limiting current densities (1.2×10^{-3} mA/cm^2 for Pt and 2×10^{-3} mA/cm^2 for Au) [20]. Quijada et al. studied the electrochemical behaviour of SO_2 with polycrystalline gold electrodes. They observed the reduction [44] and the oxidation [45] of SO_2 on gold electrodes. Regarding the oxidation of SO_2 on gold electrodes, it starts at a potential of 0.6 V, with a peak located between 0.75 and 0.85 V. This peak was higher when the concentration of SO_2 dissolved in H_2SO_4 increased, making evident that this process is mass transfer-controlled, as mentioned before, and similar as on platinum electrodes. These authors also used sulphur-modified gold electrodes from the study of Samec and Weber [43]. SO_2 was reduced on the gold electrode, and then the effect of sulphur coverage on the kinetics of the SO_2 oxidation was examined. In those conditions, the oxidation of the SO_2 peak shifts to lower potentials, for a sulphur coverage of 0.5—from that point the oxidation peak shifts back. They concluded that gold exhibits a far better performance towards SO_2 oxidation when compared to platinum. O'Brien et al. [46] compared sulphur catalysis on gold and platinum. They observed that less sulphur is formed on gold electrodes, and it does not affect the catalytic activity as much as on platinum. Allen et al. [31] observed that a gold substrate is naturally catalytically active and does not require sulphur coverage for high activity; in addition, the oxidation mechanism on gold does not change with E_{low}, whereas on platinum it does. Kriek et al. [19] modelled the electro-oxidation of SO_2 on transition metals. By calculating the maximum oxidation rate, they observed that platinum and gold are the best candidates among metals for SO_2 oxidation. They concluded that there are a limited number of metals that can be employed, due to different criteria such as inhibiting SO_2 reduction, no surface dissolution, and the metal must not be poisoned by atomic sulphur. Santasalo-Aarnio et al. [47] coated stainless steel bipolar plates with a gold layer, in order to catalyse the SO_2 oxidation and to improve the stainless steel corrosion tolerance at operation conditions. Au-coated stainless-steel bipolar plates were tested in a 100 cm^2 electrolyser for five days, and results showed that no significant loss of performance occurred. SO_2 oxidation occurs at potentials higher than 0.6 V for this electrode. Polarization curves at day one were very similar to the one at day four, indicating that the Au-coated bipolar plates have good durability and that the coating was not damaged. Nevertheless, the achieved current was very low for both days, and almost 0.5 A at 0.9 V was achieved in a 100 cm^2 single cell, which means that the current density was only around 5 mA/cm^2.

4. Palladium-Based Catalysts

Palladium, as noble metal, has been studied both pure and supported as a catalyst for SO_2 oxidation. The first studies on palladium were carried out by Lu and Ammon; in this work, the results showed a better catalytic activity for palladium than for platinum, due to a higher limiting current density for the palladium electrode [20]. They also studied the oxidation of SO_2 on palladium electrodes at different temperatures (25, 50, 70 and 90 °C), obtaining the best current density for a potential of 0.6 V (SHE) at 90 °C (1.9×10^{-3} A/cm^2). They prepared a catalyst based on palladium oxide supported on carbon, which again, gave higher current densities than a catalyst based on black platinum supported on carbon. What is more, the authors prepared a catalyst consisting of palladium oxide–titanium oxide supported on titanium, which exhibited an electrocatalytic activity quite comparable to the black-Pt/Ti.

Scott and Taama studied the oxidation of SO_2 on palladium electrodes, palladium coated graphite, and palladium-coated Ebonex (a Magneli-phase suboxide of titanium, predominantly Ti_4O_7) [26]. In this work, the palladium electrode gave a higher limiting current density than platinum. Regarding palladium-coated graphite a palladium coated Ebonex, linear sweep voltammetry shows complex curves exhibiting higher current densities than Pd, probably due to its greater exposed surface area. However, the Pd-coated electrodes showed deterioration. Colón-Mercado and Hobbs exanimated a Pd supported on carbon catalyst, which showed worse catalytic activity when compared with a catalyst based on platinum and supported on carbon. Furthermore, the Pd-based catalyst was less stable [30].

5. Catalysts Based on Other Compounds

Most of the works regarding SO_2 catalysis have been carried out with special interest on platinum, and to a lesser extent, on gold and palladium. However, aiming to develop a material with good catalytic properties and lower price, some other materials have been studied for SO_2 electro-oxidation.

Wiesener [24] carried out SO_2 electro-oxidations with catalysts based on mixtures of V_2O_5 and Al_2O_3 with different ratios. The main problem for this catalyst is its stability in acidic solutions. The optimal V/Al ratios were 1:3 and 1:6, because only a portion of vanadium was dissolved, with an amount ranging from 75 to 80% remaining in the catalyst. In general, this mixture showed worse catalytic activity than platinum. Appleby and Pinchon examined the catalytic activity of active carbons, graphites, carbon blacks, transition metal carbides, and precious metals supported on carbons [21]. Their results showed that graphites and carbides had no catalytic activity for SO_2 electro-oxidation. Active carbons had intermediate activity, but current densities were too low for use in any practical device. Lu and Ammon studied catalysts like $RuOx$-TiO_2 and $IrOx$-TiO_2, supported on titanium, ruthenium, rhenium, iridium and rhodium (Figure 7) [20]. Ru showed similar catalytic activity as platinum; however, Ir, Re and Rh electrodes were relatively inactive for SO_2 oxidation; also, $RuOx$-TiO_2/Ti and $IrOx$-$TiOx$/Ti electrodes are very ineffective for the electrochemical oxidation of SO_2 in acidic media.

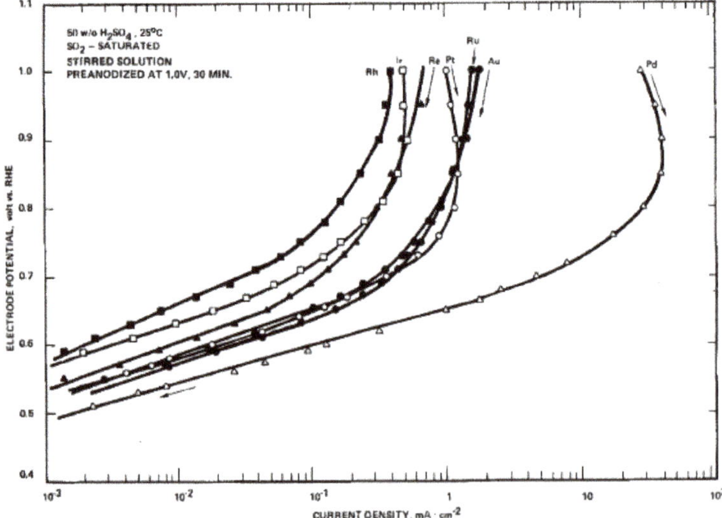

Figure 7. Tafel plots for SO_2 oxidation on smooth electrodes of Pt, Pd, Au, Ru, Re, Ir, and Rh in 50 wt % sulphuric acid at 25 °C [20]. Reproduced with permission of [12]. Copyright Electrochemical Society. 1980.

Scott and Taama also carried out voltammograms for glassy carbon, graphite electrodes, and lead oxides showing instability at high potentials [26]. Mu et al. [48] worked on the electrochemical oxidation of sulfur dioxide on nitrogen-doped graphite (NG) treated at temperatures ranging from 700 to 1000 °C. The catalytic activity of this material was compared with the activity of commercial 50% platinum supported on carbon and only Vulcan carbon XC-72. The results showed that NG treated at temperatures above 900 °C have better catalytic activity than the Vulcan carbon XC-72 without Pt, but worse than 50% Pt/C. The BET surface area was measured for the doped graphite and increased with the temperature of the thermal treatment—for example, the BET area for NG800 (thermal treatment at 800 °C) was 301 m^2/g, and for NG1000 (thermal treatment at 1000 °C) it was 425.8 mg/cm^2. Potgieter et al. [49] evaluated polycrystalline rhodium as a catalyst for SO_2 electro-oxidation. When compared with platinum, Rh showed a lower catalytic activity and was more susceptible to poisoning by adsorbed intermediate sulphur species. Similar to Pt, for Rh a decrease in starting potential resulted in an increase on the onset potential, but the catalytic activity of Rh was very limited compared with Pt, which may indicate that Rh is not suitable for SO_2 electro-oxidation. Tulskiy et al. [50] studied graphite anodes coated with different catalysts, which were Pt, MoO_3, RuO_2 and WO_3. Polarization curves showed that the catalytic activity of those materials could be arranged following the sequence $Pt > RuO_2 > MoO_3 > WO_3$. Zhao et al. [51] developed an Fe–N-Doped carbon-cladding catalyst with excellent SO_2 electrooxidation performance, close to the performance of Pt/C. It showed better stability when tested in H_2SO_4. Linear sweep voltammetry shows a catalytic activity similar to 20% Pt/C below 0.7 V (NHE). Physical characterization of the Fe–N-Doped carbon cladding showed a high surface area, mesoporous structures, and large pore volumes, which contribute to the formation of active sites and fast transport of reactants, which are beneficial for SO_2 electro-oxidation.

6. New Tendencies

The typical proton exchange membrane used in the electrolysis cell is a Nafion membrane. However, Nafion-based membranes have several limitations, including the inability to operate at elevated temperatures and decreased performance observed when exposed to high acid concentrations [35,52]. Thus, nowadays there is a tendency to work with Polybenzimidazole-based membranes (PBI), which work at high current densities to produce high sulphuric acid concentrations, and hence improve the efficiency of the electrolysis step of the cycle, as their proton conductivity does not rely on water [52]. These type of PBI-based membranes have been proposed for operating at high temperature (100–200 °C) for PEMFC technologies since 1995 [53–56]. In the case of the SO_2 electrolysis, high temperatures will have a positive impact, as the voltage losses (e.g., kinetic and ohmic resistances) would decrease [52] and the acid concentration produced could be higher [57]. Nevertheless, although the use of PBI-based membranes (sulfonated one) dates from 2012 [52], tests at temperatures higher than 90 °C were not performed. Garrick et al. [58] has recently shown results with a sulfonated PBI membrane and commercial electrodes from BASF with 1.0 $mgPt/cm^2$ for both anode and cathode in a SO_2 electrolysis cell operating at 110 °C. They concluded that the membrane resistance was not adversely affected by acid concentration, which offers benefits not seen when using Nafion. On the other hand, the large anodic overpotentials that exists in this system suggest a need for improved catalysts, and kinetics would improve with the higher temperatures afforded PBI-based membranes [58]. In this sense, higher temperatures will mean new challenges for new materials for this system, not only in terms of membranes and catalysts, but other parts of the cell. Our group has recently been working on the improvement of the Westinghouse cycle, using PBI-based membranes and novel catalyst supports for the electrochemical stage of that cycle. In the case of the catalyst support, we have proposed SiC–TiC based materials, according the previous results obtained for high temperature PEMFCs [59–62]. Recent results obtained by our group and not published yet have demonstrated that these non-carbonaceous supports can be a good candidate for the SO_2 electrolysis at high temperatures and very highly acidic conditions. Table 1 shows the values of the current at 1.0 V of cyclic voltammetry in sulphuric acid 1 M, reached before and after some electrochemical characterization tests of different catalysts. One of them is Pt supported on Vulcan carbon, and is commercial available; the other two catalysts were synthesized in our labs using the same method reported elsewhere [63], but with different support materials—in one case, the catalyst support was Vulcan XC 72, and the other was a binary carbide, SiC–TiC.

Table 1. Current values at 1.0 V of cyclic voltammetry, carried out before and after electrochemical tests of Pt-based catalysts supported on different materials.

	Intensity (A)	
	Before	**After**
Pt/C Commercial	0.31	0.29
Pt/C handmade	0.30	0.27
Pt/SiC–TiC	0.22	0.32

It can be observed that the highest currents were achieved by the catalysts based on carbon supports. After some electrochemical characterization tests (out of the scope of this manuscript) that could be considered as an accelerate degradation test, the activity of the catalysts based on carbon supports, as well as the commercial and the handmade catalysts decreased around 6.4% and 10%, respectively. On the other hand, the activity of the catalysts based on the binary carbide support increased, which means that these novel supports show a high electrochemical stability for the electrochemical oxidation of SO_2 and are very promising for this electrochemical system.

7. Conclusions

This review points out that Pt-based catalysts are the most promising materials to be implemented in the electrolyzers of the Westinghouse process. Nevertheless, their performance is still lower than the target required for full scale applications (at least 0.5 A/cm^2 at a cell voltage of 0.6 V is recommended), as can be observed in Table 2, where the most relevant catalysts reported in this work are shown. Further work has to be done in the coming years in order to reach a marketable technology.

Table 2. Most relevant results for different catalysts employed on SO_2 electrolysis.

Ref.	Year	Catalyst	Electrode	Current (A/cm^2) (@ V vs. RHE)	Pressure (bar)	Temperature (°C)	[H$_2$SO$_4$] (wt %)	Test Array
[24]	1973	V/Al$_2$O$_3$ (1:3)	6.9 mgV/Al/cm^2	0.048 (0.69)	1	100	3.8 (M)	half cell
[21]	1980	Platinum	3 mgPt/cm^2	0.316 (0.65)	1	50	55	half cell
[25]	1982	Pt/C	7 mgPt/cm^2	0.200 (0.77)	1	50	50	single cell
[26]	1999	Pd	Palladium electrode	0.033 (0.76)	1	25	0.5 (M)	half cell
[26]	1999	Graphite	Graphite electrode	0.021 (0.76)	1	70	0.5 (M)	half cell
[28]	2005	Pt/C	0.88 mgPt/cm^2	0.300 (0.73)	4	70	30	single cell
[13]	2007	Pt/C	1 mgPt/cm^2	0.400 (0.79)	1	80	(SO$_2$ gas)	single cell
[45]	2012	Au	Au electrode	0.080 (0.70)	1	22	1 (M)	half cell
[33]	2013	Pt/C	1 mgPt/cm^2	0.100 (0.75)	1	80	30	single cell
[18]	2014	Pt-Cr(1:2)/C	1 mgPt-Cr/cm^2	0.020 (0.80)	1	25	30	half cell
[48]	2015	NG	2 mg	0.15 A (0.9)	1	25	0.5 (M)	rotating disk
[47]	2016	Au	Au coating on 904L	0.005 (0.90)	1	25	15	single cell
[47]	2016	Au	Au coating on 904L	0.045 (0.75)	1	25	15	half cell
[49]	2016	Pt	Platinum electrode	0.200 (0.73)	1	25	0.5 (M)	rotating disk
[50]	2016	RuO$_2$	3 mgRuO$_2$/cm^2	0.1 (0.74)	1	25	0.5 (M)	single cell
[50]	2016	WO$_3$	3.8 mgWO$_3$/cm^2	0.1 (0.84)	1	25	0.5 (M)	single cell

As it is a general trend in the development of low-temperature fuel cells and electrolyzers, researchers try to decrease the Pt loading, in order to lower the price and make this technology more competitive. Nowadays, the typical Pt loading in the electrolyzers of the Westinghouse process is around 1 mg Pt/cm^2. However, this value may be easily decreased if novel deposition techniques and novel supports are translated from already existing PEMFC technology. Another way that is currently being used to the face decrease in the use of Pt is replacing it with other metals. Pt–Cr and Au catalysts are the most promising substitutes of Pt catalysts, although results obtained with them are currently far away from the desired targets. A final way to improve the SO_2 depolarized electrolysis for the production of hydrogen is the rise in the operation temperature, which will favor the kinetics of the process, and hence, decrease the requirements of Pt in the electrode. However, this increase must face important challenges, such as the lower cell durability because of the higher thermal stress, which affects not only the catalyst but also other cell components, such as the membranes.

Anyhow, it is important to keep in mind that most of the results found in the literature come from studies carried out in half cells or in three-electrode assemblies using rotating electrodes. More essays in complete electrolyzers are required to support the results obtained with those catalysts, as well as clarifying the more relevant aspects of the scale-up of the process. Hence, there is still a very wide slot in this technology for development, which hopefully will be filled in the next years.

Author Contributions: All the authors co-wrote the manuscript; all of them commented on the manuscript as well. S.D.-A. and J.L. conceived the structure of the review.

Funding: This research was funded by the Junta de Comunidades de Castilla-La Mancha and the FEDER—EU Program, Project ASEPHAM. Grant number "SBPLY/17/180501/000330".

Conflicts of Interest: The authors declare no conflict of interest. The funders had no role in the design of the study; in the collection, analyses, or interpretation of data; in the writing of the manuscript; or in the decision to publish the results.

References

1. Albrecht, U.; Altmann, M.; Barth, F.; Bünger, U.; Fraile, D.; Lanoix, J.-C.; Pschorr-Schoberer, E.; Vanhoudt, W.; Weindorf, W.; Zerta, M.; et al. *Study on Hydrogen from Renewable Resources in the EU*; Final Report; Ludwig-Bölkow-Systemtechnik GmbH & Hinicio S.A.: Brussels, 2015; Volume 39, ISBN 9789292461386.

2. Winter, C.J.; Nitsch, J. *Hydrogen as an Energy Carrier: Technologies, Systems, Economy*; Springer: Berlin/Heidelberg, Germany, 2012; ISBN 9783642615610.

3. Sattler, C.; Roeb, M.; Agrafiotis, C.; Thomey, D. Solar hydrogen production via sulphur based thermochemical water-splitting. *Sol. Energy* **2017**, *156*, 30–47. [CrossRef]

4. Masudy-Panah, S.; Moakhar, R.S.; Chua, C.S.; Kushwaha, A.; Dalapati, G.K. Stable and E ffi cient CuO Based Photocathode through Oxygen-Rich Composition and Au—Pd Nanostructure Incorporation for Solar-Hydrogen Production. *ACS Appl. Mater. Interfaces* **2017**. [CrossRef] [PubMed]

5. Masudy-Panah, S.; Eugene, Y.-J.K.; Khiavi, N.D.; Katal, R.; Gong, X. Aluminum-incorporated p-CuO/n-ZnO photocathode coated with nanocrystal-engineered TiO_2 protective layer for photoelectrochemical water splitting and hydrogen generation. *J. Mater. Chem. A* **2018**, *6*, 11951–11965. [CrossRef]

6. Li, X.; Yu, J.; Low, J.; Fang, Y.; Xiao, J.; Chen, X. Engineering heterogeneous semiconductors for solar water splitting. *J. Mater. Chem. A Mater. Energy Sustain.* **2014**, 1–50. [CrossRef]

7. Xinxin, W.; Kaoru, O. Thermochemical Water Splitting for Hydrogen Production Utilizing Nuclear Heat from an HTGR *. *Tsinghua Sci. Technol.* **2005**, *10*, 270–276.

8. Bhosale, R.R.; Kumar, A.; Van Den Broeke, L.J.P.; Gharbia, S.; Dardor, D.; Jilani, M.; Folady, J.; Al-fakih, M.S. Solar hydrogen production via thermochemical iron oxide-iron sulfate water splitting cycle. *Int. J. Hydrogen Energy* **2014**, *40*, 1639–1650. [CrossRef]

9. Bilgen, E. Solar hydrogen production by hybrid thermochemical processes. *Sol. Energy* **1988**, *41*, 199–206. [CrossRef]

10. Gorensek, M.B. Hybrid sulfur cycle flowsheets for hydrogen production using high-temperature gas-cooled reactors. *Int. J. Hydrogen Energy* **2011**, *36*, 12725–12741. [CrossRef]

11. Gorensek, M.B.; Edwards, T.B. Energy Efficiency Limits for a Recuperative Bayonet Sulfuric Acid Decomposition Reactor for Sulfur Cycle Thermochemical Hydrogen Production. *Ind. Eng. Chem. Res.* **2009**, *48*, 7232–7245. [CrossRef]

12. Brecher, L.E.; Wu, C.K.; Enlectrolytic decomposition of water. (Westinghouse Electric Corporation, Pittsburg, PA, USA). Personal communication, 1975.

13. Staser, J.; Ramasamy, R.P.; Sivasubramanian, P.; Weidner, J.W. Effect of Water on the Electrochemical Oxidation of Gas-Phase SO_2 in a PEM Electrolyzer for H_2 Production. *Electrochem. Solid-State Lett.* **2007**, *10*, E17–E19. [CrossRef]

14. Sivasubramanian, P.; Ramasamy, R.; Freire, F.; Holland, C.; Weidner, J. Electrochemical hydrogen production from thermochemical cycles using a proton exchange membrane electrolyzer. *Int. J. Hydrogen Energy* **2007**, *32*, 463–468. [CrossRef]

15. Leybros, J.; Saturnin, A.; Mansilla, C.; Gilardi, T.; Carles, P. Plant sizing and evaluation of hydrogen production costs from advanced processes coupled to a nuclear heat source: Part II: Hybrid-sulphur cycle. *Int. J. Hydrogen Energy* **2010**, *35*, 1019–1028. [CrossRef]

16. Jeong, Y.H.; Kazimi, M.S. Optimization of the Hybrid Sulfur Cycle for Nuclear Hydrogen Generation. *Nucl. Technol.* **2007**, *159*, 147–157. [CrossRef]

17. Lu, P.W.T. Technological aspects of sulfur dioxide depolarized electrolysis for hydrogen production. *Int. J. Hydrogen Energy* **1983**, *8*, 773–781. [CrossRef]

18. Xue, L.; Zhang, P.; Chen, S.; Wang, L. Pt-based bimetallic catalysts for SO_2-depolarized electrolysis reaction in the hybrid sulfur process. *Int. J. Hydrogen Energy* **2014**, *39*, 14196–14203. [CrossRef]

19. Kriek, R.J.; Rossmeisl, J.; Siahrostami, S.; Björketun, M.E. H_2 production through electro-oxidation of SO_2: Identifying the fundamental limitations. *Phys. Chem. Chem. Phys.* **2014**, *16*, 9572–9579. [CrossRef] [PubMed]

20. Lu, P.W.T.; Ammon, R.L. An Investigation of Electrode Materials for the Anodic Oxidation of Sulfur Dioxide in Concentrated Sulfuric Acid. *J. Electrochem. Soc.* **1980**, *127*, 2610. [CrossRef]

21. Appleby, A.J.; Pinchon, B. Electrochemical aspects of the H_2SO_4 SO_2 thermoelectrochemical cycle for hydrogen production. *Int. J. Hydrogen Energy* **1980**, *5*, 253–267. [CrossRef]

22. Seo, E.T. Determination of Sulfur Dioxide in Solution By Voltammetry. *J. Electroanal. Chem.* **1964**, *7*, 184–189.

23. Seo, E.T.; Sawyer, D.T. Electrochemical oxidation of dissolved sulphur dioxide at platinum and gold electrodes. *Electrochim. Acta* **1965**, *10*, 239–252. [CrossRef]

24. Wiesener, K. The electrochemical oxidation of sulphur dioxide at porous catalysed carbon electrodes in sulphuric acid. *Electrochim. Acta* **1973**, *18*, 185–189. [CrossRef]

25. Lu, P.W.T.; Ammon, R.L. Sulfur dioxide depolarized electrolysis for hydrogen production: Development status. *Int. J. Hydrogen Energy* **1982**, *7*, 563–575. [CrossRef]

26. Scott, K.; Taama, W.M. Investigation of anode materials in the anodic oxidation of sulphur dioxide in sulphuric acid solutions. *Electrochim. Acta* **1999**, *44*, 3421–3427. [CrossRef]

27. Weidner, J.W.; Sivasubramanian, P.; Holland, C.E.; Freire, F.J. Electrochemical Generation of Hydrogen via Gas Phase Oxidation of Sulfur Dioxide and Hydrogen Bromide Hydrogen using Thermal and/or Electrical Energy. In Proceedings of the 2005 AIChE Annual Meeting, Cincinnati, OH, USA, 30 October–4 November 2005.

28. Steimke, J.L.; Steeper, T.J. *Characterization Testing of H_2O-SO_2 Electrolyzer at Ambient Pressure*; Technical Report; U.S. Department of Energy: Washington, DC, USA, 2005. [CrossRef]

29. Report of Savannah Westinghouse Company, 2006, by Steimke, J.L.; Steeper, T.J. Available online: https://classic.ntis.gov/assets/pdf/st-on-cd/DE2006892717.pdf (accessed on 7 January 2019).

30. Colón-Mercado, H.R.; Hobbs, D.T. Catalyst evaluation for a sulfur dioxide-depolarized electrolyzer. *Electrochem. Commun.* **2007**, *9*, 2649–2653. [CrossRef]

31. Allen, J.A.; Rowe, G.; Hinkley, J.T.; Donne, S.W. Electrochemical aspects of the Hybrid Sulfur Cycle for large scale hydrogen production. *Int. J. Hydrogen Energy* **2014**, *39*, 11376–11389. [CrossRef]

32. Lee, S.K.; Kim, C.H.; Cho, W.C.; Kang, K.S.; Park, C.S.; Bae, K.K. The effect of Pt loading amount on SO_2 oxidation reaction in an SO_2-depolarized electrolyzer used in the hybrid sulfur (HyS) process. *Int. J. Hydrogen Energy* **2009**, *34*, 4701–4707. [CrossRef]

33. Xue, L.; Zhang, P.; Chen, S.; Wang, L.; Wang, J. Sensitivity study of process parameters in membrane electrode assembly preparation and SO_2 depolarized electrolysis. *Int. J. Hydrogen Energy* **2013**, *38*, 11017–11022. [CrossRef]

34. Krüger, A.J.; Krieg, H.M.; Van Der Merwe, J.; Bessarabov, D. Evaluation of MEA manufacturing parameters using EIS for SO_2 electrolysis. *Int. J. Hydrogen Energy* **2014**, *39*, 18173–18181. [CrossRef]

35. Staser, J.A.; Gorensek, M.B.; Weidner, J.W. Quantifying Individual Potential Contributions of the Hybrid Sulfur Electrolyzer. *J. Electrochem. Soc.* **2010**, *157*, B952–B958. [CrossRef]

36. Falch, A.; Lates, V.; Kriek, R.J. Combinatorial Plasma Sputtering of PtxPdy Thin Film Electrocatalysts for Aqueous SO_2 Electro-oxidation. *Electrocatalysis* **2015**, *6*, 322–330. [CrossRef]

37. Cooper, J.S.; McGinn, P.J. Combinatorial screening of fuel cell cathode catalyst compositions. *Appl. Surf. Sci.* **2007**, *254*, 662–668. [CrossRef]

38. Kleinke, M.; Knobel, M.; Bonugli, L.O.; Teschke, O. Amorphous alloys as anodic and cathodic materials for alkaline water electrolysis. *Int. J. Hydrogen Energy* **1997**, *22*, 759–762. [CrossRef]

39. Jayaraman, S.; Hillier, A.C. Construction and Reactivity Screening of a Surface Composition Gradient for Combinatorial Discovery of Electro-Oxidation Catalysts. *J. Comb. Chem.* **2004**, *6*, 27–31. [CrossRef]

40. Falch, A.; Lates, V.A.; Kotzé, H.S.; Kriek, R.J. The Effect of Rapid Thermal Annealing on Sputtered Pt and Pt3Pd2Thin Film Electrocatalysts for Aqueous SO_2 Electro-Oxidation. *Electrocatalysis* **2016**, *7*, 33–41. [CrossRef]

41. Falch, A.; Badets, V.A.; Labrugère, C.; Kriek, R.J. Co-sputtered PtxPdyAlzthin film electrocatalysts for the production of hydrogen via SO_2(aq) electro-oxidation. *Electrocatalysis* **2016**, *7*, 376–390. [CrossRef]

42. Xu, F.; Cheng, K.; Yu, Y.; Mu, S. One-pot synthesis of $Pt/CeO_2/C$ catalyst for enhancing the SO_2 electrooxidation. *Electrochim. Acta* **2017**, *229*, 253–260. [CrossRef]

43. Samec, Z.; Weber, J. Study of the Oxidation of SO_2 Dissolved 0.5 M H_2SO_4 on a Gold Electrode-I Stationary Electrode. *Electrochim. Acta* **1975**, *20*, 403–412. [CrossRef]

44. Quijada, C.; Huerta, F.J.; Morallón, E.; Vázquez, J.L.; Berlouis, L.E.A. Electrochemical behaviour of aqueous SO_2 at polycrystalline gold electrodes in acidic media: A voltammetric and in situ vibrational study. Part 1. Reduction of SO_2: Deposition of monomeric and polymeric sulphur. *Electrochim. Acta* **2000**, *45*, 1847–1862. [CrossRef]

45. Quijada, C.; Morallón, E.; Vázquez, J.L.; Berlouis, L.E.A. Electrochemical behaviour of aqueous SO_2 at polycrystalline gold electrodes in acidic media. A voltammetric and in-situ vibrational study. Part II. Oxidation of SO_2 on bare and sulphur-modified electrodes. *Electrochim. Acta* **2001**, *46*, 651–659. [CrossRef]

46. O'Brien, J.A.; Hinkley, J.T.; Donne, S.W. Electrochemical Oxidation of Aqueous Sulfur Dioxide II. Comparative Studies on Platinum and Gold Electrodes. *J. Electrochem. Soc.* **2012**, *159*, 585–593. [CrossRef]

47. Santasalo-Aarnio, A.; Lokkiluoto, A.; Virtanen, J.; Gasik, M.M. Performance of electrocatalytic gold coating on bipolar plates for SO_2 depolarized electrolyser. *J. Power Sources* **2016**, *306*, 1–7. [CrossRef]

48. Mu, C.; Hou, M.; Xiao, Y.; Zhang, H.; Hong, S.; Shao, Z. Electrochemical oxidation of sulfur dioxide on nitrogen-doped graphite in acidic media. *Electrochim. Acta* **2015**, *171*, 29–34. [CrossRef]

49. Potgieter, M.; Parrondo, J.; Ramani, V.K.; Kriek, R.J. Evaluation of Polycrystalline Platinum and Rhodium Surfaces for the Electro-Oxidation of Aqueous Sulfur Dioxide. *Electrocatalysis* **2016**, *7*, 50–59. [CrossRef]

50. Tulskiy, G.; Tulskaya, A.; Skatkov, L.; Gomozov, V.; Deribo, S. Electrochemical synthesis of hydrogen with depolarization of the anodic process. *Electrochem. Energy Technol.* **2016**, *2*, 13–16. [CrossRef]

51. Zhao, Q.; Hou, M.; Jiang, S.; Ai, J.; Zheng, L.; Shao, Z. Excellent Sulfur Dioxide Electrooxidation Performance and Good Stability on a Fe-N-Doped Carbon-Cladding Catalyst in H_2SO_4. *J. Electrochem. Soc.* **2017**, *164*, H456–H462. [CrossRef]

52. Jayakumar, J.V.; Gulledge, A.; Staser, J.A.; Kim, C.-H.; Benicewicz, B.C.; Weidner, J.W. Polybenzimidazole Membranes for Hydrogen and Sulfuric Acid Production in the Hybrid Sulfur Electrolyzer. *ECS Electrochem. Lett.* **2012**, *1*, F44–F48. [CrossRef]

53. Wainright, J.S.; Wang, J.-T.; Weng, D.; Savinell, R.F.; Litt, M. Acid-Doped Polybenzimidazoles: A New Polymer Electrolyte. *J. Electrochem. Soc.* **1995**, *142*, L121–L123. [CrossRef]

54. Lobato, J.; Cañizares, P.; Rodrigo, M.A.; Linares, J.J.; Manjavacas, G. Synthesis and characterisation of poly[2,2-(m-phenylene)-5,5-bibenzimidazole] as polymer electrolyte membrane for high temperature PEMFCs. *J. Memb. Sci.* **2006**, *280*, 351–362. [CrossRef]

55. Li, Q.; Jensen, J.O.; Savinell, R.F.; Bjerrum, N.J. High temperature proton exchange membranes based on polybenzimidazoles for fuel cells. *Prog. Polym. Sci.* **2009**, *34*, 449–477. [CrossRef]

56. Lobato, J.; Cañizares, P.; Rodrigo, M.A.; Úbeda, D.; Pinar, F.J. Promising $TiOSO_4$ Composite Polybenzimidazole-Based Membranes for High Temperature PEMFCs. *ChemSusChem* **2011**, *4*, 1489–1497. [CrossRef]

57. Steimke, J.L.; Steeper, T.J.; Colón-Mercado, H.R.; Gorensek, M.B. Development and testing of a PEM SO_2-depolarized electrolyzer and an operating method that prevents sulfur accumulation. *Int. J. Hydrogen Energy* **2015**, *40*, 13281–13294. [CrossRef]

58. Garrick, T.R.; Wilkins, C.H.; Pingitore, A.T.; Mehlhoff, J.; Gulledge, A.; Benicewicz, B.C.; Weidner, J.W. Characterizing Voltage Losses in an SO_2 Depolarized Electrolyzer Using Sulfonated Polybenzimidazole Membranes. *J. Electrochem. Soc.* **2017**, *164*, F1591–F1595. [CrossRef]

59. Lobato, J.; Zamora, H.; Cañizares, P.; Plaza, J.; Rodrigo, M.A. Microporous layer based on SiC for high temperature proton exchange membrane fuel cells. *J. Power Sources* **2015**, *288*, 288–295. [CrossRef]

60. Lobato, J.; Zamora, H.; Plaza, J.; Rodrigo, M.A. Composite Titanium Silicon Carbide as a Promising Catalyst Support for High-Temperature Proton-Exchange Membrane Fuel Cell Electrodes. *ChemCatChem* **2016**, *8*, 848–854. [CrossRef]

61. Lobato, J.; Zamora, H.; Plaza, J.; Cañizares, P.; Rodrigo, M.A. Enhancement of high temperature PEMFC stability using catalysts based on Pt supported on SiC based materials. *Appl. Catal. B Environ.* **2016**, *198*, 516–524. [CrossRef]

62. Zamora, H.; Plaza, J.; Velhac, P.; Cañizares, P.; Rodrigo, M.A.; Lobato, J. SiCTiC as catalyst support for HT-PEMFCs. Influence of Ti content. *Appl. Catal. B Environ.* **2017**, *207*, 244–254. [CrossRef]

63. Millán, M.; Zamora, H.; Rodrigo, M.A.; Lobato, J. Enhancement of Electrode Stability Using Platinum–Cobalt Nanocrystals on a Novel Composite SiCTiC Support. *ACS Appl. Mater. Interfaces* **2017**, *9*, 5927–5936. [CrossRef] [PubMed]

 *catalysts*

Article

Comparison of Direct and Mediated Electron Transfer for Bilirubin Oxidase from Myrothecium Verrucaria. Effects of Inhibitors and Temperature on the Oxygen Reduction Reaction

Riccarda Antiochia [1], Diego Oyarzun [2], Julio Sánchez [3] and Federico Tasca [3,*]

[1] Department of Chemistry and Drug Technologies, Sapienza University of Rome, 00185 Rome, Italy; riccarda.antiochia@uniroma1.it

[2] Departamento de Química, Universidad Tecnológica Metropolitana, Av. José Pedro Alessandri 1242, Santiago 7750000, Chile; diequim@gmail.com

[3] Facultad de Química y Biología, Universidad de Santiago de Chile, Santiago 9170020, Chile; julio.sanchez@usach.cl

* Correspondence: Federico.tasca@usach.cl

Received: 11 October 2019; Accepted: 28 November 2019; Published: 11 December 2019

Abstract: One of the processes most studied in bioenergetic systems in recent years is the oxygen reduction reaction (ORR). An important challenge in bioelectrochemistry is to achieve this reaction under physiological conditions. In this study, we used bilirubin oxidase (BOD) from *Myrothecium verrucaria*, a subclass of multicopper oxidases (MCOs), to catalyse the ORR to water via four electrons in physiological conditions. The active site of BOD, the T2/T3 cluster, contains three Cu atoms classified as T2, T3α, and T3β depending on their spectroscopic characteristics. A fourth Cu atom; the T1 cluster acts as a relay of electrons to the T2/T3 cluster. Graphite electrodes were modified with BOD and the direct electron transfer (DET) to the enzyme, and the mediated electron transfer (MET) using an osmium polymer (OsP) as a redox mediator, were compared. As a result, an alternative resting (AR) form was observed in the catalytic cycle of BOD. In the absence and presence of the redox mediator, the AR direct reduction occurs through the trinuclear site (TNC) via T1, specifically activated at low potentials in which T2 and T3α of the TNC are reduced and T3β is oxidized. A comparative study between the DET and MET was conducted at various pH and temperatures, considering the influence of inhibitors like H_2O_2, F^-, and Cl^-. In the presence of H_2O_2 and F^-, these bind to the TNC in a non-competitive reversible inhibition of O_2. Instead; Cl^- acts as a competitive inhibitor for the electron donor substrate and binds to the T1 site.

Keywords: oxygen reduction reaction; bilirubin oxidase; direct electron transfer; mediated electron transfer; osmium polymer

---

1. Introduction

One of the most interesting phenomena to have been intensively studied over the last 25 years is the electronic coupling between the redox cofactor of proteins and electrodes by direct (DET) or mediated electron transfer (MET) reactions [1,2]. The oxygen reduction reaction (ORR) is considered one of the most important electrocatalytic reactions due to its implication in biological systems as well as in industrial processes [3,4]. Multicellular living organisms use oxygen as an electron (e−) acceptor, for example, in the respiratory chain [5,6]. In fuel cells chemical energy is converted into electrical energy through the ORR [3,4,7–9]. Many fungi and bacteria use oxygen as a signalling molecule [10–12] to form hydrogen peroxide and degrade wood and lignocelluloses. In nature this reaction is catalysed mainly by Cu containing proteins, such as multicopper oxidases (MCO) [13,14]. MCOs constitute a family of enzymes that includes laccase (Lc), ascorbate oxidase, bilirubin oxidase (BOD), CotA, Fet3p, CueO and ceruloplasmin [15]. They are important for Fe metabolism and related disease states, antibiotic biosynthesis, biotechnology, bioremediation, biosensing and BFCs [8,15]. These enzymes can be immobilized directly on the electrode (DET) or using small redox molecules (MET) that facilitate electron transfer between the redox centre of the enzyme and the electrode [1,2,6,16]. Several studies have been conducted on Cu enzymes to gather more information about the ORR and how to reproduce this reaction with synthetic and biological materials. In this sense, MCOs have been identified as particularly efficient catalysts of the ORR, carrying out fast, four-electron reduction at low overpotentials [17]. BODs, a sub-class of MCOs containing four Cu ions per one enzyme molecule, were discovered in 1981 by Tanaka and Murao [18], and first used for the detection of bilirubin [19] and later for the reduction of O_2 [20,21]. According to their magnetic and optical properties, the Cu atoms are classified into three types T1, T2 and a binuclear T3 (T2 and T3 sites are combined in a tri-nuclear cluster or TNC) [16,22–24]. The single-electron oxidation of substrates occurs at the proximal T1 site, with the 4-electron reduction of O_2 taking place at the TNC [19]. The 4 electrons are quickly transferred from the T1 site to the MCO TNC via a histidine-cysteine-histidine ligand [19]. In 2008 the resting oxidised form of the enzyme was proposed and lately an AR form was characterized by crystallography, spectroscopy and electrochemistry [25,26]. AR was found to occur only with high potential MCO where the T1 has a high enough potential to perform the one-electron oxidation of the TNC that produces AR [27]. The activation for catalysis of these different forms was assessed. The results showed that the AR form can only be activated by low-potential reduction, in contrast to the resting oxidized form, which was activated via T1 at high potential [26]. This difference in activity was correlated to differences in redox states of the two forms [26].

A main limitation for enzyme-modified electrodes is related to the electron transfer between the enzyme active sites and solid conducting supports (electrodes). Alternatively, mediated electron transfer (MET) involves small redox molecules that shuttle electrons from enzyme to electrode [5,28]. Most of the highest reported current densities are produced by MET-type electrodes, because enzyme molecules may be electrochemically active in multiple layers and at any orientation [29]. Osmium polymers have been used extensively as electron-transfer mediators to bind MCOs to the electrode surface [30–39]. Nevertheless, the reactivation of BOD in the presence of osmium polymer has never been studied before. Herein, the electron transfer processes for the different states of BOD in the presence of an osmium polymer were studied and compared to DET.

It is known that the activity of Lc and BOD can be inhibited by halide anions; however, BOD is less sensitive to Cl^- ions than Lc [23,40]. For example, F^- acts as a non-competitive inhibitor that interrupts electron transfer between MCOs' T1 and TNC [23,41]. Other studies on Lc with H_2O_2 have demonstrated a reversible non-competitive inhibition of the enzyme due to oxidation of the T3 Cu site [31]. To the best of our knowledge, the effect of inhibitors was never studied in the presence of redox mediators and compared to DET. In this work the effect of pH, temperature and of the inhibitors was studied when in the presence or in the absence of osmium polymer and the action from H_2O_2 and F^- anions were analysed in the activation of the AR form in the DET and MET of bilirubin oxidase on graphite electrodes.

2. Results and Discussion

2.1. Electrochemical Characterization

MCOs couple the oxidation of bilirubin to the reduction of oxygen to water through four copper atoms: a T1 Cu atom that functions as an electron relay and a trinuclear cluster (2T3 denominated as T3α, T3β and 1T2 Cu atoms) where the reduction occurs. The substrate for T1 can be substituted with a source of electrons, for example, a graphite electrode.

Therefore, when BOD is immobilized on the surface of graphite electrodes the T1 redox centre is responsible for shuttling the electrons from the electron surface to the T2/T3 cluster.

The cyclic voltammograms provide important information about the immobilized enzyme. It is important to note that electrocatalysis by adsorbed enzymes presents a residual slope and therefore voltammetry curves show a linear response, suggesting that the electrode follows Ohm's law. However, Armstrong and co-workers explain that this effect could be because the enzymes are not adsorbed homogeneously on the electrode [42,43]. At the same time, this effect is much more evident when the studies were carried out at different pH and temperature as will be described below. Meanwhile, spectroscopies and electrochemistry are used to identify different resting forms of MCOs. One of these forms corresponds to a fully oxidized resting form (RO), which is derived from the decay of the native intermediate (NI) [27]. Another resting state called alternative resting form (AR) was identified in some high redox potential MCOs, such as BOD from CotA, *Bacillus subtilis* [44], *Myrothecium verrucaria* [45] (Mv-BOD), *Trachyderma tsunodae* [45] (Tt-BOD), *Magnaporthae orizae* [45] (Mo-BOD), *Pleurotos* ostreatus and *Bacillus pumilus* [45] (Bp-BOD). Figure 1A shows typical cyclic voltammograms (CVs) of an edge graphite electrode modified with 10 µL of 0.2 mM BOD solution in the absence and in the presence of O_2. The scans were carried out cathodically in 0.1 M phosphate buffer solution at pH 7. In the absence of O_2 (blue line), non-faradaic processes were observed due to the large surface area of the graphite electrode and its corresponding charging/discharging process. A non-modified graphite electrode would present an almost identical cyclic voltammogram, and therefore it is not shown in the figure. The black and red lines in Figure 1A correspond to the first and second scans, respectively, in O_2 saturated buffer for a fresh electrode. For the first scan (Figure 1A, black line) the voltammogram on the forward sweep is characterized by two hysteresis observed at ~0.40 V and ~0.1 V vs SCE, indicating an activation step at low potentials for the ORR. During the second scan the onset for O_2 reduction occurs at ~0.5 V. The hysteresis obtained during the first scan suggests that many molecules of the enzyme in direct electron contact with the graphite electrode are not catalytically active in a redox state corresponding to the AR form of the enzyme [27,46,47]. Therefore, the AR form can be activated for O_2 catalysis, at low potential for the Cu sites that form the TNC. In the purification process of BOD from *Myrothecium verrucaria* it is common to use NaCl [48], which could cause the formation of the AR. These results are in agreement with Poulpiquet and co-workers, as they show that an activation step at low potentials for O_2 reduction corresponds to the partially reduced AR form in which TNC T2 and T3α are reduced and T3β is oxidized [47]. Furthermore, the same group recently demonstrated that the full reduction of the AR form of Lc cannot occur via T1, but only via T3β [47]. A DET process through the TNC supposes that the TNC is sufficiently close to the interface to exchange electrons. The insert of Figure 1A shows a magnification of the potential zone of 0.4 V to observe the hysteresis of the first scan for BOD activation. In the second cycle (red line) the onset for the ORR occurs at ~0.5 V potential without presenting any hysteresis, indicating the activation of the enzyme. The osmium redox polymers allow an efficient electron transfer by wiring multiple layers of the immobilized molecules and promoting stable adsorption [30]. A high redox potential OsP ($E^{0'}$~0.21 V vs. SCE) was selected from a previously reported library of Os polymers [49]. Figure 1B depicts the electrochemical response of a freshly prepared BOD-OsP modified graphite electrode in 0.1 M phosphate buffer solution, at pH 7.0 in the absence and in the presence of O_2, and at a scan rate of 5 mVs^{-1} the scans swept from 0.8 V to −0.2 V. In the absence of O_2 (black line), the characteristic redox wave of the OsP with an $E^{0'}$ ~0.21 V for Os^{III}/Os^{II} redox couple is clearly defined. In the presence of O_2 (red line first scan, blue line second scan) the first scan of the voltammogram shows two hysteresis at ~0.32 V and ~ −0.05 V, suggesting again that the T2/T3 centre for the ORR is activated at low potentials, while during the second scan the onset for O_2 reduction is seen at 0.40 V. When the potential values of the hysteresis under DET and MET are compared (from Figure 1A,B), a displacement of ~40 mV may be seen in the cathodic direction for T1 and T2/T3. Most probably this displacement is caused by the reduction of the OsP, which should be followed by the reduction of the Cu atoms. The ligands of the OsP could affect the BOD sites to make the ORR take place or the AR form could be activated only at sufficiently low potentials for O_2 reduction. Similar current densities and onsets for the ORR of the BOD-OsP electrodes have been obtained before [33], but this is the first time that activation of BOD enzyme has been reported in the presence of the OsP hydrogel. The activation of the T2/T3 redox centre could be the result of the reduction of the centre by the T1 cluster or directly by the electrode or by electrons that flow from OsP redox sites. Solomon and co-workers [27] showed that when T3 are partially oxidized and T2 is fully oxidized (AR form of the enzyme), the T2/T3 sites have a significantly lower electron-affinity than the high potential T1 Cu. The fact that the T2/T3 reduction occurs at lower potentials in the presence of the OsP could indicate that the reaction occurs principally through the OsP atoms, and therefore reducing the Os atoms is necessary before the electrons are transferred to the T2/T3 couple [27].

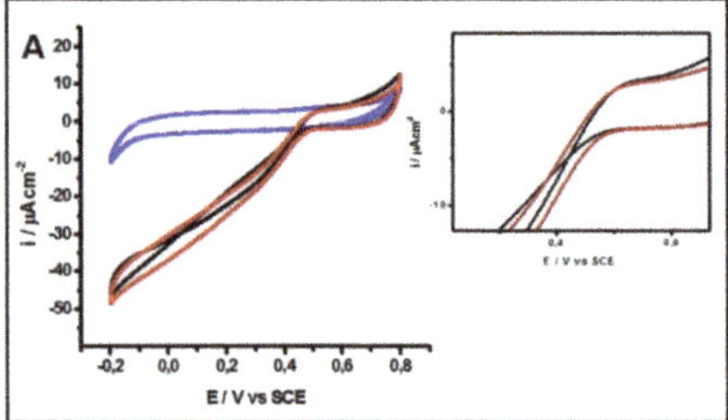
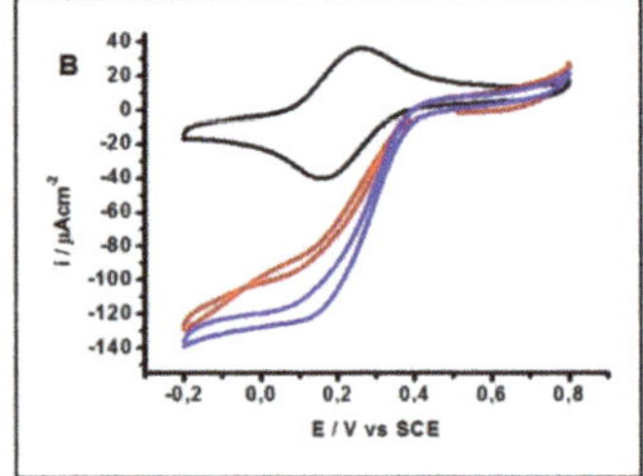

Figure 1. (**A**) Cyclic voltammogram of a BOD modified electrode. Insert: zoom on the 0.4 V to observe the activation zone of the second cycle (black line). (**B**) Cyclic voltammogram of the Os/Poly/BOD/ modified electrode. Conditions: 0.1 M phosphate buffer solution, pH 7.0 in absence (blue line) and in presence (black line first scan, red line second scan) of O_2. Scan rate: 5 mV s^{-1}.

2.2. Effect of pH

In solution, *mv*BOD is active in a wide pH range from pH 5 to pH 8 [45,50,51]. Also it has been proven that *mv*BOD is very stable in the presence and in the absence of Os polymer or other mediator, losing around 10% of activity during one day under operating conditions [45,50–52]. The effect of pH on the MCO catalytic mechanism has only been partly elucidated, but it is well resumed in [45,53]. Figure 2A shows cyclic voltammograms using a BOD modified electrode in phosphate buffer 0.1 M in the presence of O_2 at different pH (pH 8 black line, pH 7 red line, pH 6 green line, and pH 5 blue line). Phosphate buffer was used to obtain various pHs because it is known that it does not cause interference with the catalytic activity of the enzyme. The capacity of the buffer in the aforementioned range is of 0.02 moles and therefore it can be used for the purpose. The shape of the curves is the same for pH 5, 6 and 7 in the first scan, where two hystereses can be observed at different potential values that, in all cases shift to more negative potentials as the pH is increased. Hysteresis is consistent with the presence of the AR form of the enzyme on the surface of the electrode and with the low-potential half wave that corresponds to the T2/T3 reduction, indicating a proton dependent reaction [27,46,47]. The reduction of the T2/T3 sites follows ~60 mV dependence per pH units. During the second scan (after reduction of the T2/T3 cluster) the low-potential half wave disappears. The high potential wave corresponds to the T1 site and does not change from the first to the second scan but it follows ~30 mV dependence per pH unit. At pH 8, to observe the hysteresis at low potential half wave is very difficult and during the second scan very low activity could be measured. This is in agreement with the studies of dos Santos [42], which indicates that at this pH the enzyme is not active. In acidic conditions the intramolecular processes are fast: the current rate is determined by the T1 redox cycling but the T1 potential does not define the catalytic activity. At neutral or alkaline pH, the intramolecular processes are slow and determine the current rate as well as the T1 potential defines the catalytic activity. Furthermore, the current decreases at alkaline pH due to the lack of proton transfer and electrochemical driving force as suggested by do Santos and co-workers [42]. For the second scan, the onset for the ORR depends on pH, moving to more positive potentials as the pH decreases. Figure 2B shows the voltammograms corresponding to an electrode modified with osmium polymer, in phosphate buffer 0.1 M in the presence of O_2 at different pH (pH 8 black line, pH 7 red line, pH 6 green line, and pH 5 blue line). Table 1 summarizes the potential of the high and low potential hysteresis for oxygen reduction in DET and MET, at pH 7. The first scan is characterized by two hysteresis at different potential values at pH 5. In this case, as the pH increases the potential values displace to more negative potentials. However, at pH 6 and 7 the low wave potential is not clearly defined, probably because the ET from the OsP to the TNC is so fast that this process is not observed. The hysteresis at low wave potential is consistent with the presence of an AR form of the enzyme after reductive activation below this hysteresis; this AR form is converted into the RO form of the enzyme confirmed through the second cycle of the voltammogram. At pH 8 hysteresis is not observed and only the wave for the Os^{III}/Os^{II} redox couple is clearly defined with E′ ~0.27 V. The first and the second scan yielded the same results in practice, evidencing the inactivation of the enzyme at this pH. It is important to note that the current density for hysteresis at low potential half wave increases when the OsP is present compared to the DET, because the ET is favoured from the enzyme to the electrode by the Os atoms present in high concentration. Likewise, hysteresis at the low potential half wave for MET shifts to negative values due to the OsP redox potential (the redox mediator should be in the reduced form to donate electrons to the cluster) improving the ET from the electrode to OsP and from OsP to the T2/T3 cluster. It is noteworthy that under physiological conditions, these four electrons are exchanged sequentially through the T1 copper site of the enzyme and further transferred by internal electron transfer (ET) to the trinuclear T2/T3 cluster.

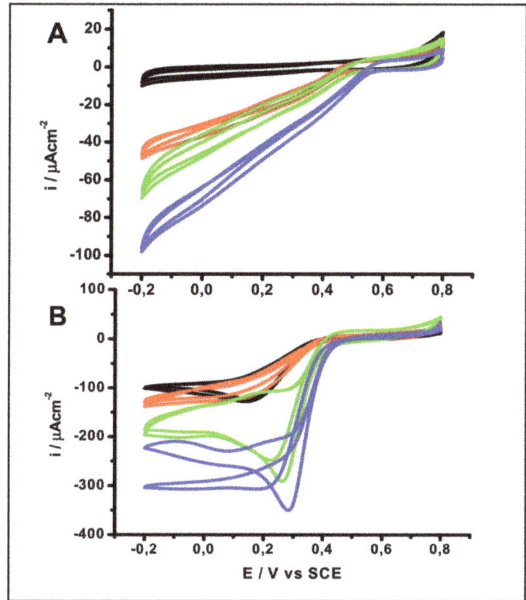

Figure 2. Cyclic voltammetry of (**A**) BOD and (**B**) Os/Poly/BOD/PEGDGE at various pHs. Conditions: phosphate buffer 0.1 M, at various pHs: pH 8 (black), pH 7 (red), pH 6 (green), pH 5 (blue) Scan rate: 5 mV s⁻¹.

Table 1. Summary of the potential corresponding to the high potential and low potential hystereses for oxygen reduction in DET and MET, at pH 7.

	DET		MET	
pH	High Potential Hysteresis, V	Low Potential Hysteresis, V	High Potential Hysteresis, V	Low Potential Hysteresis, V
7	0.440 ± 5	0.100 ± 5	0.300 ± 5	0.000 ± 5

2.3. Temperature Effect

The oxygen reduction catalysis is affected by various factors, among which is temperature [38]. *mv*BOD in solution is fully active from 30 °C to 60 °C whereas it loses 50% of original activity at 20 °C [6]. The effect of the temperature on the catalytic activity of BOD in DET and MET systems in the ORR was studied. Figure 3A shows cyclic voltammograms for O_2 reduction at different temperatures, all at 5 mV s^{-1} using a BOD electrode in phosphate buffer 0.1 M and in presence of O_2. The studies were performed in a temperature range from 10 °C to 50 °C. Figure 3A shows cyclic voltammograms in the presence of O_2 at 10 °C (black line) and 50 °C (green line). During the first scan, at both temperatures hysteresis close to 0.4 V and 0.35 V, respectively, were observed, suggesting the presence of the AR form. At 50 °C the activity is 10% lower than at 30 °C. Figure 3B shows similar experiments to Figure 3A but in the presence of OsP. At 10 °C (black line) and at 20 °C (results not shown), hysteresis appears near 0.30 V during the first scan for BOD activation. At 30 °C (red line) and 50 °C (green line) a different behaviour is seen. Not clear hysteresis can be observed and differences between the catalytic current and the shape of the voltammograms (~10% of catalytic current) of the first and second scan are minor. Probably due to the swelling of the polymer, the electron transfer at these temperatures is so fast that the enzyme is activated very fast and is ready for oxygen reduction. It should be noted that at 50 °C the enzyme presents a higher activity than at 10 °C, which is in contrast with the results showed in Figure 3A, where no major difference appears at different temperatures. These results can be explained due to a faster interaction between the enzymatic redox sites and the OsP.

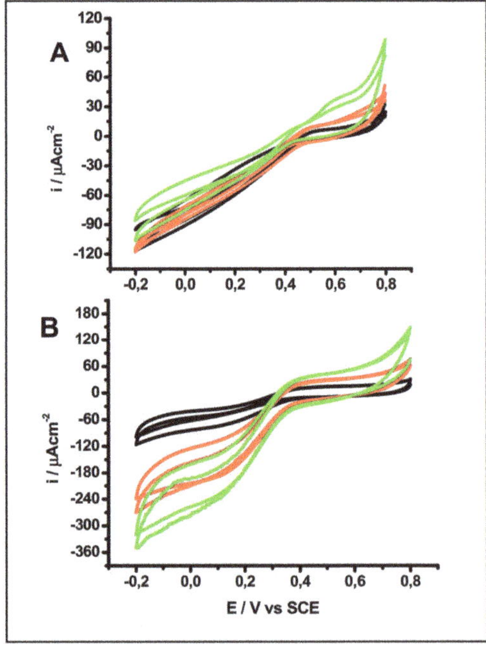

Figure 3. Cyclic voltammetry of (**A**) BOD and (**B**) Os/Poly/BOD/PEGDGE at various temperatures. Conditions: phosphate buffer 0.1 M, pH 7, at different temperature: 10 °C (black), 30 °C (red), 50 °C (green). Scan rate: 5 mV s^{-1}.

2.4. Inhibitory Effect

In 2010, Calvo et al. [31] reported the first evidence of the inhibitor effect of H_2O_2 on MCOs. Milton and Minteer have reported the reversible inhibition of Lc by H_2O_2 under DET and MET with ABTS (2,2-azino bis(3-ethylbenzothiazoline-6-sulfonic acid) [23,54].

Figure 4A shows chronoamperometry measurements of an activated BOD modified electrode polarized at E = 0 V in the presence of O_2 and during the additions of 1 μM, 5 μM and 10 μM H_2O_2. Clearly, the presence of 1 μM of H_2O_2 negatively affects the electrocatalytic current and the subsequent additions of H_2O_2 (5 μM and 10 μM final concentrations) confirm the inhibitory process. Figure 4B shows cyclic voltammetric measurements of the same electrode employed during the chronoamperometry of Figure 4A in the absence of H_2O_2. The second scan (black line) clearly shows that an activation process takes place after the first scan (reductive process, red line) but hysteresis is not evident, as opposed to the previous voltammograms. The current density obtained after reduction was ~−41 μAcm^{-2}. Therefore, 71% of the activity was recovered. Figure 4B shows the effect of H_2O_2 on a BOD-OsP under the same conditions shown in Figure 4A. After the first addition of 1 μM H_2O_2 no change in the current density was observed. However, when H_2O_2 was added at final concentrations of 10 μM, a small decrease of current density was seen (~8% compared to the constant current of the chronoamperometry). Therefore, only high concentrations (<10 μM) of H_2O_2 can inhibit the electrocatalytic process in the presence of OSP. In order to reactivate the process the system needs to be saturated with O_2, so it can displace the peroxide located in the cluster. The voltammogram corresponding to the enzyme activation is illustrated in Figure 4D. In this case the current density reaches −239 μAcm^{-2} at 0 V. Therefore, when compared with the constant current of chronoamperometry measurements, it represents 80% of the recovery in enzyme activity. In both cases the results for DET and MET indicated that the enzyme showed a non-competitive reversible inhibition of O_2 by exogenous H_2O_2, which interacted with the TNC [31,32]. Under DET and using a BDO cathode, Milton and co-workers [23] suggested that H_2O_2 follows a non-competitive inhibition. These same authors reported the inhibition of laccase with H_2O_2 under both conditions: DET and MET with 2,2′-azino-bis(3-ethylbenzothiazoline-6-sulfonic acid) (ABTS) as redox mediator. They indicated that laccase under DET follows an uncompetitive inhibition by H_2O_2 binding to the T2/T3 cluster [54]. Calvo showed the inhibition of laccase by H_2O_2 using an multilayer osmium derivative poly (allylamine) redox mediator and found that exogenous H_2O_2 inhibits laccase due to T3 Cu^{+1} [31].

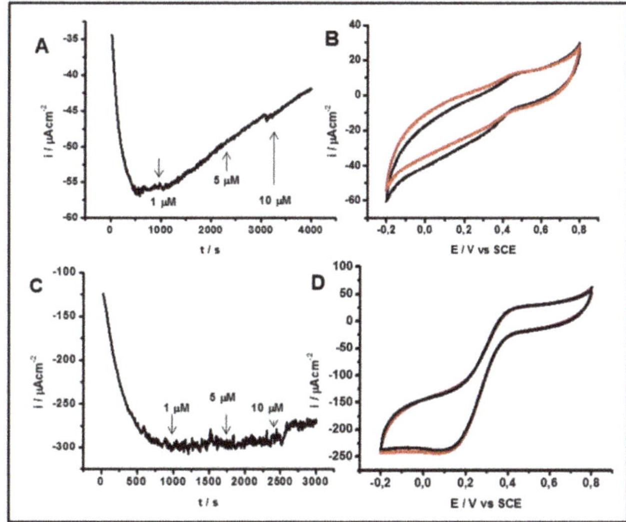

Figure 4. Chronoamperometry of (**A**) BOD and (**C**) Os/Poly/BOD/PEGDGE. Conditions: phosphate buffer 0.1 M, pH 7, at E = 0 V (vs SCE) in presence of O_2 and with H_2O_2 addition at different concentrations 1 μM, 5 μM and 10 μM. Later to the H_2O_2 addition the cyclic voltammetry of (**B**) BOD and (**D**) Os/Poly/BOD/PEGDGE was recorded. Scan rate: 5 mV s^{-1}.

BODs can be inhibited by halide anions such as F^- and Cl^- [23]. Figure 5A shows the chronoamperometric of a BOD electrode in the presence of O_2 at pH 7 at various NaF concentrations (1 µM, 5 µM and 10 µM). In DET experiments concentrations up to 10 mM were added where a considerable current density loss was observed for the ORR. Figure 5B shows cyclic voltammetric measurements in the absence of F^- of the same electrode used in Figure 5A. Once more, the reactivation process is not very clear and there are not many differences between the first and second scan, indicating that the enzyme keeps its activated form and the inhibition process is not caused by the oxidization/reduction process to the TNC, because O_2 is not able to completely displace the fluoride ions from the active site. In Figure 5C chronoamperometric measurements are shown in the presence of an OsP at different NaF concentrations (100 µM, 1000 µM and 10000 µM). In this case only very high concentrations of NaF (>1 mM) would affect the electrocatalytic process. These results indicate that fluoride acts as a non-competitive inhibitor, binding to the T2/T3 cluster of the enzyme, interrupting the ET between T1 and the TNC, and finally preventing the ORR. EPR studies revealed that F^- interacts with TNC only in the presence of the oxidized T3 Cu (II) [55]. This would explain why a bigger inhibition in the redox mediator electrode is observed, since F^- is blocking the active site favoured by the OsP. Figure 5D shows cyclic voltammetry of the BOD reactivation at pH 7 without inhibitor, where it is possible to displace F^-, recovering 95% of the activity. These results are in agreement with those published by other authors, which suggest that F^- ions inhibit DET and MET [55–57].

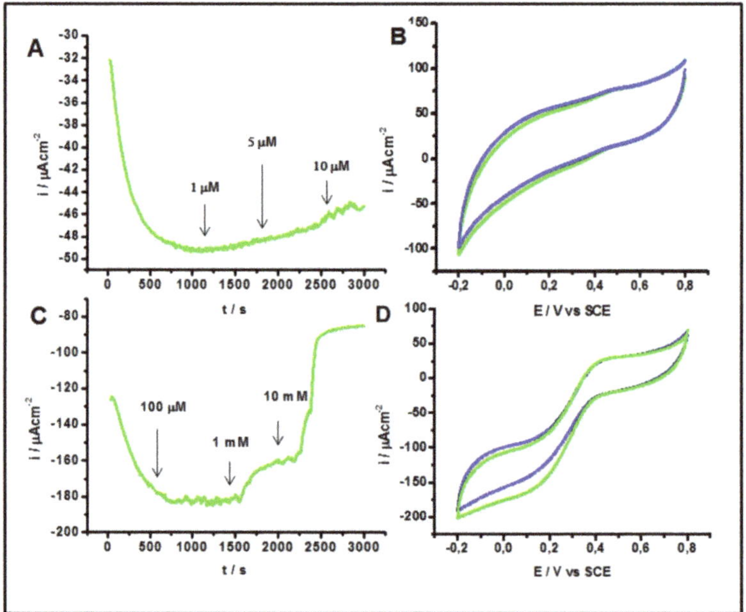

Figure 5. Chronoamperometry of (**A**) BOD and (**C**) Os/Poly/BOD/PEGDGE. Conditions: phosphate buffer 0.1 M, pH 7, at E= 0 V (vs SCE) in presence of O_2 and with NaF addition. The concentrations of NaF added were 1 µM, 5 µM and 10 µM for the BOD and 100 µM, 1000 µM and 10,000 µM for the Os/Poly/BOD/PEGDGE electrode, respectively. Later to the NaF addition the cyclic voltammetry of (**B**) BOD and (**D**) Os/Poly/BOD/PEGDGE was recorded. Scan rate: 5 mV s^{-1}.

Another halide that inhibits the BOD is Cl^-. Figure 6A shows the chronoamperometric measurements for a BOD modified electrode in the presence of O_2 at different NaCl concentrations (1 mM, 0.01 M and 0.1 M). This figure evidences a decrease in the current density when Cl^- concentration >0.1 M is added. Figure 6B shows the cyclic voltammetry in a fresh solution in the absence of Cl^- and in the presence of O_2 using the same electrode used for the chronoamperometry shown in Figure 6A. When the BOD is reactivated O_2 displaces the Cl^- located in the T1 site because no activation in the second cycle (brown line) can be observed. Figure 6C shows the effect of Cl^- on a BOD electrode modified with an OsP under the same conditions indicated above, in which a slight decrease in the current (4%) can be appreciated when the maximum concentration of Cl^- (0.1 M) is added. The voltammograms in the absence of Cl^- (using the same electrode in Figure 6C) are shown in Figure 6D. This figure shows catalytic activity that suggests that Cl^- act as a competitive inhibitor with respect to the electron donor substrate, and that O_2 displaces the Cl^- present in the T1 site. These results indicate that the enzyme can show a reversible inhibition behaviour led by chloride ions so Cl^- blocks the ET to the T1 site, instead of binding it to the TNC [58]. However, Jensen and co-workers [59] have witnessed the Cl^- inhibition effect on laccase in DET. Nevertheless, high resistance to Cl^- inhibition under DET has been reported in nanostructured electrodes, which suggests that they avoid Cl^- entry to the T1 site [56,57].

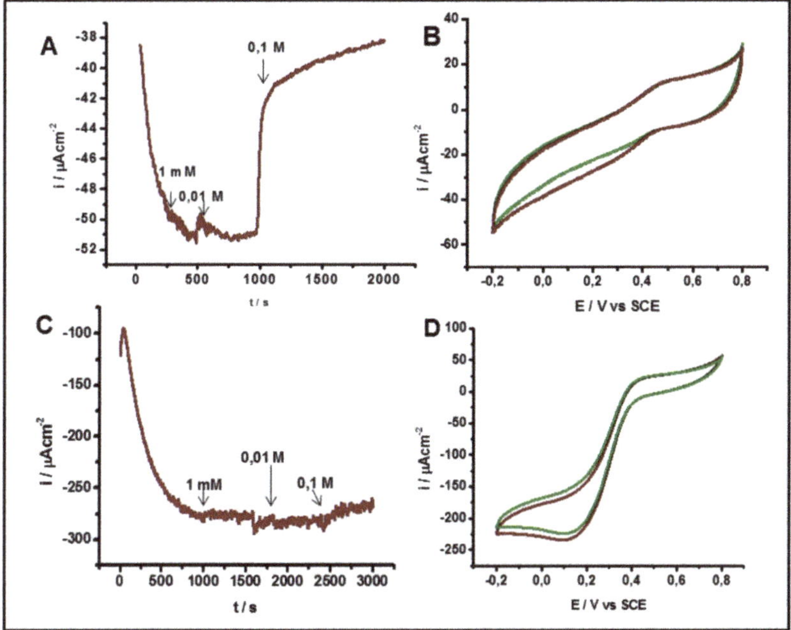

Figure 6. Chronoamperometry of (**A**) BOD and (**C**) Os/Poly/BOD/PEGDGE. Conditions: phosphate buffer 0.1 M, pH 7, at E = 0 V (vs SCE) in presence of O_2 with NaCl addition at different concentrations 1000 µM, 10,000 µM and 100,000 µM. Cyclic voltammetry of (**B**) BOD and (**D**) Os/Poly/BOD/PEGDGE was recorded after the NaCl addition. Scan rate: 5 mV s^{-1}.

3. Experimental

3.1. Chemicals

K_2HPO_4, KH_2PO_4, 30% H_2O_2, NaF, NaCl and Poly (ethylene glycol) diglycidyl ether (PEGDGE), 2,2′-Azino-bis(3-ethylbenzothiazoline-6-sulfonic acid) (ABTS) and dithionite were purchased from Sigma Aldrich (San Luis, MO, USA) with ACS purity grade. High purity (90% or higher concentration of pure enzyme) *Myrothecium verrucaria* BOD "Amano 3" was donated by Amano Enzyme Inc., (Madeline, LN, USA). The received lyophilized powder was stored at −20 °C and 0.2 mM enzyme solution were fresh prepared. The activity of the enzyme solution was checked by monitoring ABTS oxidation at 420 nm ($\varepsilon = 36$ mM^{-1}cm^{-1}) over time with stirring in 0.1 M sodium phosphate buffer. The activity test was performed on each prepared batch and resulted to be 12 ± 2 U/mg whereas one unit is defined as the amount of enzyme that oxidizes 1 μmol of ABTS per minute. The Osmium polymer $[Os(bpy)_2(PVI)_{10}Cl]^{2+/+}$ was synthetized at the National University of Ireland, Galway following procedures already reported in [60–63]. The average molecular weight of the polymer as determined by viscometry in ethanol was 130,000 g/mol [64]. By taking the mean value of the anodic and cathodic peak potentials of the CV, the E°-value of the Os polymer was found to be +180 mV vs. SCE (sat. KCl). The polymer is positively charged so to increases the solubility in water and the electrostatic interaction with anionic regions of the enzyme [64].

3.2. Preparation of BOD and BOD-OsP Modified Electrodes

Edge-plane pyrolytic graphite electrodes of 5 mm diameter were purchased from PINE (Durham, NC, USA). To prepare the electrodes, the end of a rod was wet polished with an in-house made electrode polisher and waterproof 800 and 1200 grit emery sandpaper and then sonicated in deionized water for 10 min to remove any residual impurities. This process also insures a highly reproducible real surface area with less than 10% deviation in the charging/discharging current of blank electrodes measured during cyclic voltammetry experiments [65,66]. The electrodes were then rinsed with Milli-Q water and air-dried. Later, 10 μL of enzyme solution (~0.2 mM) were placed on top of the polished rod and adsorption was allowed to occur. The electrodes were then stored overnight at 4 °C under controlled humidity. After 24 h electrodes were rinsed with Milli-Q water and after being inserted in a Teflon holder, they were ready to use as working electrodes. For the MET experiments, electrodes were modified through the same previous treatment (mechanical cleaning) and drop adsorption, depositing first 10 μL of the OsP dissolved in water (10 mg/mL) [49,60]. This solution was mixed with 20 μL of a BOD 0.2 mM solution and 5 μL of PEGDGE 10 mg/mL. The electrodes were kept overnight at 4 °C under controlled humidity for complete cross-linking.

3.3. Electrochemical Measurements

The electrochemical measurements were carried out in a conventional three-electrode cell. The electrodes used were a platinum spiral wire as counter electrode and a saturated calomel electrode as reference electrode. All the results are related to this reference electrode. Current densities were calculated based on the geometrical area of the working electrode. The experiments were conducted with at least three repetitions. A 0.1 M phosphate buffer (K_2HPO_4, KH_2PO_4) was used as electrolytic solution. The range of pH effect was studied between 5 to 8. The inhibitors used were H_2O_2, NaF and NaCl added at different concentrations. Electrochemical experiments were performed with an AUTOLAB PGSTAT204 potensiostat (Utrecht, Netherlands).

4. Conclusions

In this study, MvBOD was studied in the absence and in the presence of OsP. The influence of pH and temperature, as well as the effects of H_2O_2, Cl^-, and F^- inhibitors were evaluated. Particular attention was paid to the activation process of the enzyme and therefore to the reduction of the AR form. The ORR can be activated only if the potential of the enzyme is reduced at values lower than 0.1 V, where TNC's T2 and T3α are reduced and T3β is oxidized. In the presence of the OsP a shift of ~40 mV in the cathodic direction for the T1 and for the T2/T3 reduction can be observed probably because the OsP needs to be reduced before the electrons can be transferred to the Cu sites. The ORR process was studied at various pHs. Both DET and MET showed very low activity at pH 8. Under DET the T2/T3 reduction showed a proton dependent reaction [27,46,47] that follows ~ 60 mV dependence per pH units. Instead, T1 reduction follows ~ 30 mV dependence per pH unit. Under MET the T1 reduction process still follows ~ 30 mV dependence per pH unit, while the reduction of the trinuclear cluster at pH 5 and 6 cannot be determined. At 30 °C and higher temperatures, in the presence of OsP, hysteresis, which enhances the activation of the enzyme, was not evident. In both DET and MET, in the presence of H_2O_2, a non-competitive reversible inhibition occurred, suggesting that it binds to the TNC of the enzyme. In MET, the inhibition occurs only at very high concentrations (>10 μM of H_2O_2).

The inhibition by halides showed a different mechanism for F^- and Cl^-. F^- acts as a non-competitive inhibitor, binding to the oxidized T3 Cu (II) [55]. In the case of F^-, under DET, small concentrations inhibited the ORR, while during MET concentrations higher than 1 mM this was necessary. After using a fresh buffer solution no hysteresis were observed. Cl^- showed a competitive inhibition [56,57]. Nevertheless, the inhibition process occurred only during DET and at very high concentrations of NaCl (>0.1 M). Using fresh buffer, a small reactivation process took place, which indicates that Cl^- could cause the AR form of the enzyme. During MET the inhibition process was less evident, suggesting that the presence of the OsP prevents the interaction between Cl^- and the T1 site.

Author Contributions: Data curation, J.S.; Formal analysis, R.A.; Investigation, D.O.; Writing—original draft, F.T.

Funding: This research was funded by Fondecyt Project 1181840.

Acknowledgments: F.T. thanks for financial support the Fondecyt Project 1181840, and Proyecto Basale Dicyt.

Conflicts of Interest: The authors declare no conflict of interest.

References

1. Balzani, V. (Ed.) *Electron. Transfer in Chemistry*; WILEY-VCH Verlag GmbH & Co. KGaA: Weinheim, Germany, 2001.

2. Gray, H.B.; Winkler, J.R. *Electron. Transfer in Metalloproteins*; Wiley-VCH: Weinheim, Germany, 2001.

3. Gamella, M.; Koushanpour, A.; Katz, E. Biofuel cells—Activation of micro- and macro-electronic devices. *Bioelectrochemistry* **2018**, *119*, 33–42. [CrossRef] [PubMed]

4. Solomon, E.I.; Stahl, S.S. Introduction: Oxygen Reduction and Activation in Catalysis. *Chem. Rev.* **2018**, *118*, 2299–2301. [CrossRef] [PubMed]

5. Gewirth, A.A.; Thorum, M.S. Electroreduction of Dioxygen for Fuel-Cell Applications: Materials and Challenges. *Inorg. Chem.* **2010**, *49*, 3557–3566. [CrossRef] [PubMed]

6. Tasca, F.; Farias, D.; Castro, C.; Acuna-Rougier, C.; Antiochia, R. Bilirubin Oxidase from Myrothecium verrucaria Physically Absorbed on Graphite Electrodes. Insights into the Alternative Resting Form and the Sources of Activity Loss. *PLoS ONE* **2015**, *10*, e0132181. [CrossRef]

7. Venegas, R.; Recio, F.J.; Riquelme, J.; Neira, K.; Marco, J.F.; Ponce, I.; Zagal, J.H.; Tasca, F. Biomimetic reduction of O_2 in an acid medium on iron phthalocyanines axially coordinated to pyridine anchored on carbon nanotubes. *J. Mater. Chem. A* **2017**, *5*, 12054–12059. [CrossRef]

8. Santoro, C.; Babanova, S.; Erable, B.; Schuler, A.; Atanassov, P. Bilirubin oxidase based enzymatic air-breathing cathode: Operation under pristine and contaminated conditions. *Bioelectrochemistry* **2016**, *108*, 1–7. [CrossRef]

9. Venegas, R.; Recio, F.J.; Zuniga, C.; Viera, M.; Oyarzun, M.-P.; Silva, N.; Neira, K.; Marco, J.F.; Zagal, J.H.; Tasca, F. Comparison of the catalytic activity for O2 reduction of Fe and Co MN4 adsorbed on graphite electrodes and on carbon nanotubes. *Phys. Chem. Chem. Phys.* **2017**, *19*, 20441–20450. [CrossRef]

10. Majumdar, S.; Lukk, T.; Solbiati, J.O.; Bauer, S.; Nair, S.K.; Cronan, J.E.; Gerlt, J.A. Roles of Small Laccases from Streptomyces in Lignin Degradation. *Biochemistry* **2014**, *53*, 4047–4058. [CrossRef]

11. Augustine, A.J.; Kragh, M.E.; Sarangi, R.; Fujii, S.; Liboiron, B.D.; Stoj, C.S.; Kosman, D.J.; Hodgson, K.O.; Hedman, B.; Solomon, E.I. Spectroscopic Studies of Perturbed T1 Cu Sites in the Multicopper Oxidases Saccharomyces cerevisiae Fet3p and Rhus vernicifera Laccase: Allosteric Coupling between the T1 and Trinuclear Cu Sites. *Biochemistry* **2008**, *47*, 2036–2045. [CrossRef]

12. Stone, J.R.; Yang, S. Hydrogen Peroxide: A Signaling Messenger. *Antioxid. Redox Signal.* **2006**, *8*, 243–270. [CrossRef]

13. Johnson, D.L.; Thompson, J.L.; Brinkmann, S.M.; Schuller, K.A.; Martin, L.L. Electrochemical Characterization of Purified Rhus vernicifera Laccase: Voltammetric Evidence for a Sequential Four-Electron Transfer. *Biochemistry* **2003**, *42*, 10229–10237. [CrossRef]

14. Filip, J.; Tkac, J. The pH dependence of the cathodic peak potential of the active sites in bilirubin oxidase. *Bioelectrochemistry* **2014**, *96*, 14–20. [CrossRef] [PubMed]

15. Solomon, E.I.; Sundaram, U.M.; Machonkin, T.E. Multicopper Oxidases and Oxygenases. *Chem. Rev.* **1996**, *96*, 2563–2606. [CrossRef] [PubMed]

16. Mano, N.; Edembe, L. Bilirubin oxidases in bioelectrochemistry: Features and recent findings. *Biosens. Bioelectron.* **2013**, *50*, 478–485. [CrossRef] [PubMed]

17. Cracknell, J.A.; Blanford, C.F. Developing the mechanism of dioxygen reduction catalyzed by multicopper oxidases using protein film electrochemistry. *Chem. Sci.* **2012**, *3*, 1567–1581. [CrossRef]

18. Murao, S.; Tanaka, N. A new enzyme "bilirubin oxidase" produced by Myrothecium verrucaria MT-1. *Agric. Biol. Chem.* **1981**, *45*, 2383–2384. [CrossRef]

19. Goldfinch, M.E.; Maguire, G.A. Investigation of the use of bilirubin oxidase to measure the apparent unbound bilirubin concentration in human plasma. *Ann. Clin. Biochem.* **1988**, *25*, 73–77. [CrossRef]

20. Tsujimura, S.; Fujita, M.; Tatsumi, H.; Kano, K.; Ikeda, T. Bioelectrocatalysis-based dihydrogen/dioxygen fuel cell operating at physiological pH. *Phys. Chem. Chem. Phys.* **2001**, *3*, 1331–1335. [CrossRef]

21. Palmore, G.T.R.; Kim, H.-H. Electro-enzymatic reduction of dioxygen to water in the cathode compartment of a biofuel cell. *J. Electroanal. Chem.* **1999**, *464*, 110–117. [CrossRef]

22. Mano, N.; Kim, H.-H.; Heller, A. On the Relationship between the Characteristics of Bilirubin Oxidases and O2 Cathodes Based on Their "Wiring". *J. Phys. Chem. B* **2002**, *106*, 8842–8848. [CrossRef]

23. Milton, R.D.; Giroud, F.; Thumser, A.E.; Minteer, S.D.; Slade, R.C.T. Bilirubin oxidase bioelectrocatalytic cathodes: The impact of hydrogen peroxide. *Chem. Commun. (Cambridge, UK)* **2014**, *50*, 94–96. [CrossRef] [PubMed]

24. Sakurai, T.; Kataoka, K. Basic and applied features of multicopper oxidases, CueO, bilirubin oxidase, and laccase. *Chem. Rec.* **2007**, *7*, 220–229. [CrossRef] [PubMed]

25. Solomon, E.I.; Augustine, A.J.; Yoon, J. O_2 Reduction to H_2O by the multicopper oxidases. *Dalton Trans.* **2008**, 3921–3932. [CrossRef] [PubMed]

26. Kjaergaard, C.H.; Durand, F.; Tasca, F.; Qayyum, M.F.; Kauffmann, B.; Gounel, S.; Suraniti, E.; Hodgson, B.; Hedman, K.O.; Mano, N.; et al. Spectroscopic and Crystallographic Characterization of "Alternative Resting" and "Resting Oxidized" Enzyme Forms of Bilirubin Oxidase: Implications for Activity and Electrochemical Behavior of Multicopper Oxidases. *J. Am. Chem. Soc.* **2012**, *134*, 5548–5551. [CrossRef] [PubMed]

27. Kjaergaard, C.H.; Jones, S.M.; Gounel, S.; Mano, N.; Solomon, E.I. Two-Electron Reduction versus One-Electron Oxidation of the Type 3 Pair in the Multicopper Oxidases. *J. Am. Chem. Soc.* **2015**, *137*, 8783–8794. [CrossRef] [PubMed]

28. Barton, S.C.; Gallaway, J.; Atanassov, P. Enzymatic biofuel cells for implantable and microscale devices. *Chem. Rev. (Washington, DC, USA)* **2004**, *104*, 4867–4886. [CrossRef]

29. Gallaway, J.W.; Calabrese Barton, S.A. Effect of redox polymer synthesis on the performance of a mediated laccase oxygen cathode. *J. Electroanal. Chem.* **2009**, *626*, 149–155. [CrossRef]

30. Szamocki, R.; Flexer, V.; Levin, L.; Forchiasin, F.; Calvo, E.J. Oxygen cathode based on a layer-by-layer self-assembled laccase and osmium redox mediator. *Electrochim. Acta* **2009**, *54*, 1970–1977. [CrossRef]

31. Grattieri, M.; Scodeller, P.; Adam, C.; Calvo, E.J. Non-competitive reversible inhibition of laccase by H2O2 in osmium mediated layer-by-layer multilayer O_2 biocathodes. *J. Electrochem. Soc.* **2015**, *162*, G82–G86. [CrossRef]

32. Scodeller, P.; Carballo, R.; Szamocki, R.; Levin, L.; Forchiassin, F.; Calvo, E.J. Layer-by-layer self-assembled osmium polymer-mediated laccase oxygen cathodes for biofuel cells: The role of hydrogen peroxide. *J. Am. Chem. Soc.* **2010**, *132*, 11132–11140. [CrossRef]

33. Ackermann, Y.; Guschin, D.A.; Eckhard, K.; Shleev, S.; Schuhmann, W. Design of a bioelectrocatalytic electrode interface for oxygen reduction in biofuel cells based on a specifically adapted Os-complex containing redox polymer with entrapped Trametes hirsuta laccase. *Electrochem. Commun.* **2010**, *12*, 640–643. [CrossRef]

34. Jenkins, P.A.; Boland, S.; Kavanagh, P.; Leech, D. Evaluation of performance and stability of biocatalytic redox films constructed with different copper oxygenases and osmium-based redox polymers. *Bioelectrochemistry* **2009**, *76*, 162–168. [CrossRef] [PubMed]

35. Lalaoui, N.; Reuillard, B.; Philouze, C.; Holzinger, M.; Cosnier, S.; Le Goff, A. Osmium (II) Complexes Bearing Chelating N-Heterocyclic Carbene and Pyrene-Modified Ligands: Surface Electrochemistry and Electron Transfer Mediation of Oxygen Reduction by Multicopper Enzymes. *Organometallics* **2016**, *35*, 2987–2992. [CrossRef]

36. Shin, H.; Cho, S.; Heller, A.; Kang, C. Stabilization of a Bilirubin Oxidase-Wiring Redox Polymer by Quaternization and Characteristics of the Resulting O_2 Cathode. *J. Electrochem. Soc.* **2009**, *156*, F87–F92. [CrossRef]

37. Shin, H.; Kang, C. Co-electrodeposition of bilirubin oxidase with redox polymer through ligand substitution for use as an oxygen reduction cathode. *Bull. Korean Chem. Soc.* **2010**, *31*, 3118–3122. [CrossRef]

38. Mano, N.; Kim, H.-H.; Zhang, Y.; Heller, A. An Oxygen Cathode Operating in a Physiological Solution. *J. Am. Chem. Soc.* **2002**, *124*, 6480–6486. [CrossRef] [PubMed]

39. Poeller, S.; Beyl, Y.; Vivekananthan, J.; Guschin, D.A.; Schuhmann, W. A new synthesis route for Os-complex modified redox polymers for potential biofuel cell applications. *Bioelectrochemistry* **2002**, *87*, 178–184. [CrossRef]

40. Wang, X.; Falk, M.; Ortiz, R.; Matsumura, H.; Bobacka, J.; Ludwig, R.; Bergelin, M.; Gorton, L.; Shleev, S. Mediatorless sugar/oxygen enzymatic fuel cells based on gold nanoparticle-modified electrodes. *Biosens. Bioelectron.* **2012**, *31*, 219–225. [CrossRef]

41. Salaj-Kosla, U.; Poller, S.; Schuhmann, W.; Shleev, S.; Magner, E. Direct electron transfer of Trametes hirsuta laccase adsorbed at unmodified nanoporous gold electrodes. *Bioelectrochemistry* **2013**, *91*, 15–20. [CrossRef]

42. Leger, C.; Jones, A.K.; Albracht, S.P.J.; Armstrong, F.A. Effect of a Dispersion of Interfacial Electron Transfer Rates on Steady State Catalytic Electron Transport in [NiFe]-hydrogenase and Other Enzymes. *J. Phys. Chem. B* **2002**, *106*, 13058–13063. [CrossRef]

43. Dos Santos, L.; Climent, V.; Blanford, C.F.; Armstrong, F.A. Mechanistic studies of the 'blue' Cu enzyme, bilirubin oxidase, as a highly efficient electrocatalyst for the oxygen reduction reaction. *Phys. Chem. Chem. Phys.* **2010**, *12*, 13962–13974. [CrossRef] [PubMed]

44. Reiss, R.; Ihssen, J.; Thony-Meyer, L. Bacillus pumilus laccase: A heat stable enzyme with a wide substrate spectrum. *BMC Biotechnol.* **2011**, *11*, 9. [CrossRef] [PubMed]

45. Mano, N.; de Poulpiquet, A. O_2 Reduction in Enzymatic Biofuel Cells. *Chem. Rev. (Washington, DC, USA)* **2018**, *118*, 2392–2468. [CrossRef] [PubMed]

46. Gentil, S.; Carriere, M.; Cosnier, S.; Gounel, S.; Mano, N.; Le Goff, A. Direct electrochemistry of bilirubin oxidase from Magnaporthe orizae on covalently-functionalized MWCNT for the design of high-performance oxygen-reducing biocathodes. *Chem. Eur. J.* **2018**, *24*, 8404–8408. [CrossRef] [PubMed]

47. De Poulpiquet, A.; Kjaergaard, C.H.; Rouhana, J.; Mazurenko, I.; Infossi, P.; Gounel, S.; Gadiou, R.; Giudici-Orticoni, M.T.; Solomon, E.I.; Mano, N.; et al. Mechanism of chloride inhibition of bilirubin oxidases and its dependence on potential and pH. *ACS Catal.* **2017**, *7*, 3916–3923. [CrossRef] [PubMed]

48. Han, X.; Zhao, M.; Lu, L.; Liu, Y. Purification, characterization and decolorization of bilirubin oxidase from Myrothecium verrucaria 3.2190. *Fungal Biol.* **2012**, *116*, 863–871. [CrossRef]

49. Zafar, M.N.; Tasca, F.; Boland, S.; Kujawa, M.; Patel, I.; Peterbauer, C.K.; Leech, D.; Gorton, L. Wiring of pyranose dehydrogenase with osmium polymers of different redox potentials. *Bioelectrochemistry* **2010**, *80*, 38–42. [CrossRef]

50. Mano, N. Features and applications of bilirubin oxidases. *Appl. Microbiol. Biotechnol.* **2012**, *96*, 301–307. [CrossRef]

51. De Poulpiquet, A.; Ciaccafava, A.; Gadiou, R.; Gounel, S.; Giudici-Orticoni, M.T.; Mano, N.; Lojou, E. Design of a H_2/O_2 biofuel cell based on thermostable enzymes. *Electrochem. Commun.* **2014**, *42*, 72–74. [CrossRef]

52. Chen, X.; Yin, L.; Lv, J.; Gross, A.J.; Le, M.; Gutierrez, N.G.; Li, Y.; Jeerapan, I.; Giroud, F.; Berezovska, A.; et al. Stretchable and Flexible Buckypaper-Based Lactate Biofuel Cell for Wearable Electronics. *Adv. Funct. Mater.* **2019**, *29*, 1905785. [CrossRef]

53. Kang, C.; Shin, H.; Heller, A. On the stability of the "wired" bilirubin oxidase oxygen cathode in serum. *Bioelectrochemistry* **2006**, *68*, 22–26. [CrossRef] [PubMed]

54. Milton, R.D.; Minteer, S.D. Investigating the reversible inhibition model of laccase by hydrogen peroxide for bioelectrocatalytic applications. *J. Electrochem. Soc.* **2014**, *161*, H3011–H3014. [CrossRef]

55. Di Bari, C.; Mano, N.; Shleev, S.; Pita, M.; De Lacey, A.L. Halides inhibition of multicopper oxidases studied by FTIR spectroelectrochemistry using azide as an active infrared probe. *JBIC J. Biol. Inorg. Chem.* **2017**, *22*, 1179–1186. [CrossRef] [PubMed]

56. Vaz-Dominguez, C.; Campuzano, S.; Ruediger, O.; Pita, M.; Gorbacheva, M.; Shleev, S.; Fernandez, V.M.; De Lacey, A.L. Laccase electrode for direct electrocatalytic reduction of O_2 to H_2O with high-operational stability and resistance to chloride inhibition. *Biosens. Bioelectron.* **2008**, *24*, 531–537. [CrossRef]

57. Tominaga, M.; Sasaki, A.; Togami, M. Bioelectrocatalytic oxygen reaction and chloride inhibition resistance of laccase immobilized on single-walled carbon nanotube and carbon paper electrodes. *Electrochemistry (Tokyo, Jpn.)* **2016**, *84*, 315–318. [CrossRef]

58. Enaud, E.; Trovaslet, M.; Naveau, F.; Decristoforo, A.; Bizet, S.; Vanhulle, S.; Jolivalt, C. Laccase chloride inhibition reduction by an anthraquinonic substrate. *Enzym. Microb. Technol.* **2011**, *49*, 517–525. [CrossRef]
59. Jensen, U.B.; Vagin, M.; Koroleva, O.; Sutherland, D.S.; Besenbacher, F.; Ferapontova, E.E. Activation of laccase bioelectrocatalysis of O_2 reduction to H_2O by carbon nanoparticles. *J. Electroanal. Chem.* **2012**, *667*, 11–18. [CrossRef]
60. Boland, S. Integrating Enzymes, Electron Transfer Mediators and Mesostructured Surfaces to Provide Bio-Electroresponsive Systems. Ph.D. Thesis, National University of Ireland, Galway, Ireland, 2008.
61. Zafar, M.N.; Tasca, F.; Gorton, L.; Patridge, E.V.; Ferry, J.G.; Noll, G. Tryptophan Repressor-Binding Proteins from Escherichia coli and Archaeoglobus fulgidus as New Catalysts for 1,4-Dihydronicotinamide Adenine Dinucleotide-Dependent Amperometric Biosensors and Biofuel Cells. *Anal. Chem.* **2009**, *81*, 4082–4088. [CrossRef]
62. Antiochia, R.; Tasca, F.; Mannina, L. Osmium-Polymer modified carbon nanotube paste electrode for detection of sucrose and fructose. *Mater. Sci. Appl.* **2013**, *4*, 15–22. [CrossRef]
63. Kurbanoglu, S.; Zafar, M.N.; Tasca, F.; Aslam, I.; Spadiut, O.; Leech, D.; Haltrich, D.; Gorton, L. Amperometric Flow Injection Analysis of Glucose and Galactose Based on Engineered Pyranose 2-Oxidases and Osmium Polymers for Biosensor Applications. *Electroanalysis* **2018**, *30*, 1496–1504. [CrossRef]
64. Heller, A. Electrical connection of enzyme redox centers to electrodes. *J. Phys. Chem.* **1992**, *96*, 3579–3587. [CrossRef]
65. Recio, F.J.; Canete, P.; Tasca, F.; Linares-Flores, C.; Zagal, J.H. Tuning the Fe (II)/(I) formal potential of the FeN4 catalysts adsorbed on graphite electrodes to the reversible potential of the reaction for maximum activity: Hydrazine oxidation. *Electrochem. Commun.* **2013**, *30*, 34–37. [CrossRef]
66. Tasca, F.; Recio, F.J.; Venegas, R.; Geraldo, D.A.; Sancy, M.; Zagal, J.H. Linear versus volcano correlations for the electrocatalytic oxidation of hydrazine on graphite electrodes modified with MN4 macrocyclic complexes. *Electrochim. Acta* **2014**, *140*, 314–319. [CrossRef]

Article

Electrochemical Oxidation of Urea on NiCu Alloy Nanoparticles Decorated Carbon Nanofibers

Ahmed Abutaleb

Department of Chemical Engineering, Jazan University, Jazan 45142, Saudi Arabia; azabutaleb@jazanu.edu.sa or engahmedabutaleb@gmail.com; Tel.: +966597997716, Fax: +966-17-3232900

Received: 6 April 2019; Accepted: 26 April 2019; Published: 28 April 2019

Abstract: Bimetallic $Cu_{3.8}Ni$ alloy nanoparticles (NPs)-anchored carbon nanofibers (composite NFs) were synthesized using a simple electrospinning machine. XRD, SEM, TEM, and TGA were employed to examine the physiochemical characteristics of these composite NFs. The characterization techniques proved that $Cu_{3.8}Ni$ alloy NPs-anchored carbon NFs were successfully fabricated. Urea oxidation (UO) processes as a source of hydrogen and electrical energy were investigated using the fabricated composite NFs. The corresponding onset potential of UO and the oxidation current density (OCD) were measured via cyclic voltammetry as 380 mV versus Ag/AgCl electrode and 98 mA/cm^2, respectively. Kinetic study indicated that the electrochemical oxidation of urea followed the diffusion controlled process and the reaction order is 0.5 with respect to urea concentration. The diffusion coefficient of urea using the introduced electrocatalyst was found to be 6.04×10^{-3} cm^2/s. Additionally, the composite NFs showed steady state stability for 900 s using chronoamperometry test.

Keywords: electrospinning; $Cu_{3.8}Ni$-nanoalloy; carbon nanofibers (NFs); urea oxidation; fuel cells

---

1. Introduction

Many valuable characteristics are detected during the application of urea in industrial fields, including its low cost, solid-state at room temperature, high chemical stability, nonflammability, nontoxicity and easy handling and transportation [1–5]. Human and animal urine contains urea as the main constituent [4,6–8]. Discharging 75 kg of polluted urea wastewater results in the production of 0.75 kg of urea [6,9]. Urea in wastewater can be converted into toxic substances such as ammonia [2,6]. However, urea can act as hydrogen source material with high hydrogen content (10 wt%) [2,6]. Compressed hydrogen, liquid hydrogen and urea can produce equivalent energy density values of 5.6, 10.1 and 16.9 MJ/L, respectively [6,7,10]. Accordingly, urea-rich wastewaters are deemed an efficient energy source through the following reactions [8,11–15]:

$$\text{Anode: } CO(NH_2)_2 + 6OH^- \rightarrow N_2 + 5H_2O + CO_2 + 6e \quad E_0 = -0.746 \text{ V} \tag{1}$$

$$\text{Cathode: } 2H_2O + O_2 + 4e \rightarrow 4OH^- \quad E_0 = +0.40 \text{ V} \tag{2}$$

$$\text{Overall: } 2CO(NH_2)_2 + 3O_2 \rightarrow 2N_2 + 4H_2O + 2CO_2 \quad E_0 = +1.146 \text{ V} \tag{3}$$

The lowered electrocatalytic output of anode material in urea fuel cells can hinder their commercialization [2,11]. Platinum-based electrocatalysts exhibit poor electrocatalytic performance despite being widely applied in many reactions [12]. On the other hand, better performance has been achieved using nickel electrodes [5,10,11,16–18]. While introduced nickel electrocatalysts have exhibited good catalytic activity for anodic reactions, these activities remain too limited for market introduction. To improve the catalytic behaviors of nickel-based electrocatalysts, two protocols can be employed. The first involves preparing nickel of different forms such as Ni nanoribbons [9], Ni nanoparticles [19], Ni nanowires [5], and Ni on carbon sponge [8] to increase the surface area of electrocatalysts. Second, nickel can be alloyed with other transition metals such as manganese [6], cobalt [18] and zinc [4].

Nanofibrous materials have recently been produced via electrospinning with enhanced mass and electron transfer [20–30]. Barakat et al. [31] have exhibited the good performance of NiO electrospun NFs electrocatalyst during methanol oxidation reactions. Abdelkareem and coauthors [2] have also prepared Ni-Cd nanoparticles on electrospun carbon nanofibers (NFs) with good electrocatalytic activity for UO. Reduced overpotential and high oxidation current density (OCD) levels were achieved from increased active surface areas using adjustable dimensional frameworks of nanocarbon support (e.g., carbon nanotubes and carbon sheets). This may support effective electrocatalytic activities during several organic materials oxidation reactions [6,11,21]. Carbon has been extensively used in many applications involving batteries and sensors due to its electrical conductivity and higher current output relative to amorphous carbon [1].

Copper has been investigated as a catalyst and cocatalyst in various chemical and electrochemical reactions [21,23,26]. To our knowledge, little work has been examined effects of the incorporation of copper, nickel, and carbon as electrocatalysts for urea fuel cells. Thus, in this work, Ni-Cu nanoparticles embedded on carbon NFs were fabricated via electrospinning to oxidize urea in alkaline solution. The introduced NFs present low onset potential and good current density (CD) values through UO. Moreover, chronoamperometry tests reveal its stable performance after 900 s.

2. Results and Discussion

2.1. Characterization

SEM images of electrospun nanofibrous mats containing NiAc, CuAc, and PVA after vacuum drying are shown in Figure 1A,B. Random distributed and smooth surfaced NFs were obtained. On the other hand, after performing the calcination step, as it is illustrated in Figure 1C, the structure shows multilayered nanofibers with entanglement network structures without interconnections of specific directions. Irregular nanoparticles are grown on the NFs surface (Figure 1C,D). Surprisingly, the nanofibrous morphology was not affected considerably by the calcination conditions. Furthermore, Elemental-EDX analysis (Figure 2) indicated the presence of Ni, Cu, and C only in the calcined NFs.

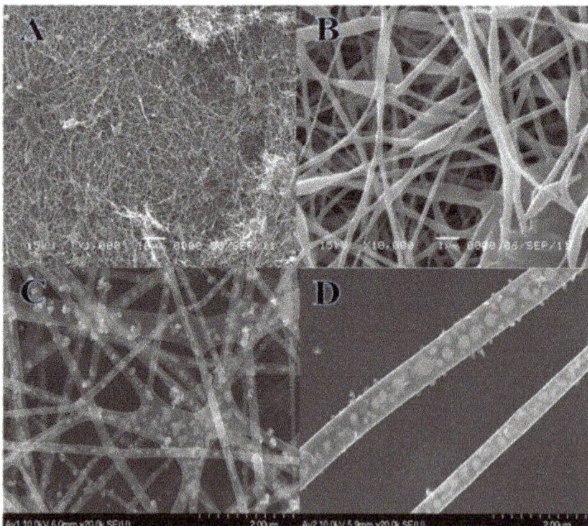

Figure 1. SEM images of electrospun nanofiber mats after drying at 60 °C for 1 day (**A**) at low magnification, (**B**) at high magnification and (**C** and **D**) FE-SEM images of electrospun nanofiber mats after sintering at 800 °C for 5 h in an Argon atmosphere.

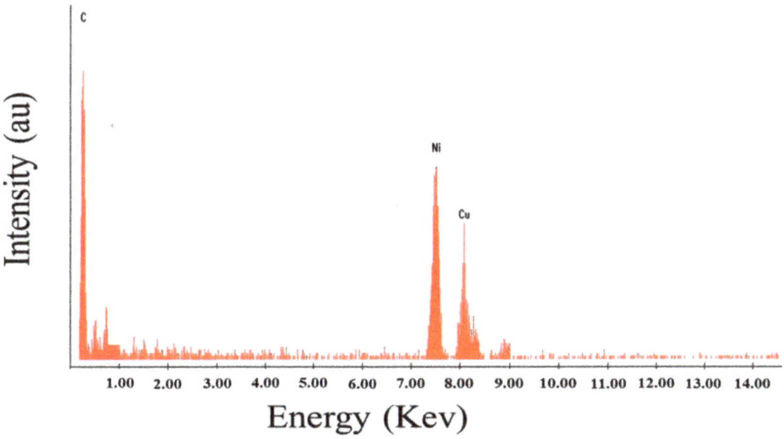

Figure 2. Elemental EDX analysis of the produced nanofibers (NFs).

Figure 3A presents typical XRD patterns of the calcined NFs. Nickel and copper are neighbors in the periodic table, so their crystal parameters are close. Therefore, the identification peaks in the XRD database are overlapped. Moreover, the produced NiCu bimetallic alloys have also different crystal structures compared to the pristine metals. For instance, nickel can be assigned in the XRD pattern if peaks are observed at 2θ values of 44.5°, 51.8° and 76.4° corresponding to (111), (200), and (220) crystal plans, respectively (JCDPS #04-0850). Interestingly, both of pure copper and $Cu_{3.8}Ni$ are identified with the same crystal plans and at close 2θ values; numerically, 43.1°, 50.8°, and 74.1°, and 43.5°, 50.8° and 74.7°, respectively (JCDPS card number 04-0836 for Cu and 09-0205 for $Cu_{3.8}Ni$). Considering that each of the two metals has melting point higher than the utilized calcination temperature; 1455 and 1085 °C for Ni and Cu, respectively, vaporization of the formed metals is not expected. Moreover, Ni precursor has a big amount compared to Cu in the initial electrospun solution, so the observed peaks in the XRD pattern (Figure 3A) can be assigned to free Ni and $Cu_{3.8}Ni$ alloy. It is noteworthy mentioning that the pristine nickel and the metallic alloy can combine in the same nanoparticle which was proved by TEM EDX analysis below. An extra peak is shown in the spectra at a 2θ value of 25.6° for the graphite-like-carbon structure. It is important to note that Ni and NiCu exhibit strong catalytic activity during the graphitization of the employed polymer, elucidating the detection of carbon [27].

The carbon content of the prepared NFs is determined via TGA analysis in an oxygen atmosphere in Figure 3B. The remaining residuals were calculated as 52.3wt% with respect to the original weight of the fabricated NFs. During calcination, metal NPs were converted in an oxygen atmosphere at high temperatures into metal oxides (NiO and CuO) while CO and/or CO_2 gases formed from carbon. Accordingly, carbon disappeared while residue powder was composed of only CuO and NiO. The carbon content of the calcined NFs was measured as 41.08 wt%. It is noteworthy mentioning that incomplete oxidation of the metals is expected during the TGA analysis under oxygen atmosphere which leads to form chemically unstable compounds. However, as it is shown in the figure, a trivial decrease in the weight was observed after T ~ 450 °C. This indicates that unstable compounds were formed, might be Cu_2O or other Cu or Ni oxides, and these compounds gradually changed to the stable Cu and Ni oxide forms; NiO and CuO. Moreover, due to the very high surface area of the investigated nanofibers, utilizing oxygen atmosphere rather than air, and performing the analysis at high temperature, full oxidation of carbon is expected. Figure 4 shows Raman Spectroscopy analysis of the produced powder. The figure illustrates the chemistry of the formed carbon with two peaks centered at roughly 1330 (D band) and 1590 cm^{-1} (G band). This shows that the PVA was transformed into carbon with a fraction of disordered sp2 C–C bonding. The first peak is found in all carbon-like carbons while the second corresponds to planar vibrations of carbon atoms in the carbon-like materials.

The increase in the I_D/I_G intensity ratio suggested the presence of six-fold rings and some defects on the surface of carbon nanofibers [28,32,33]. This result supports the XRD data, showing that the produced NFs contained carbon.

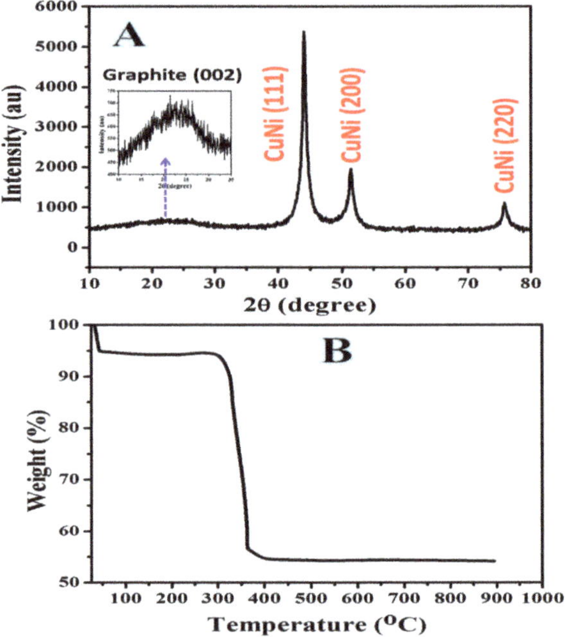

Figure 3. (**A**) XRD patterns of the calcined CuAC/NiAC/PVA mat NFs after sintering at 800 °C for 5 h in an argon atmosphere, (**B**) TGA analysis result of CuNi/carbon nanofibers under oxygen atmosphere.

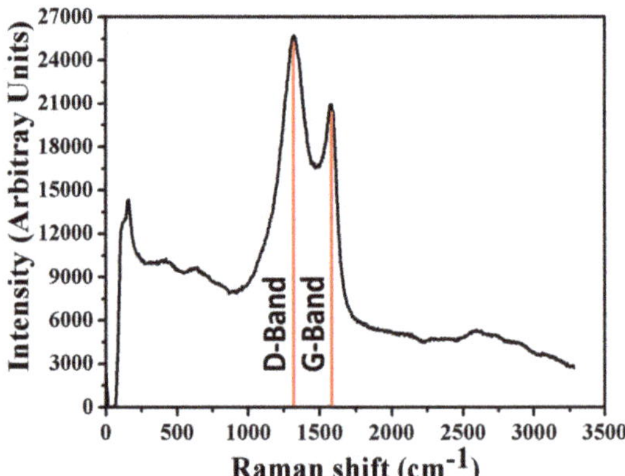

Figure 4. Raman spectrum of the prepared nanofibers.

TEM images shown in Figure 5 present the distribution of metallic NPs on the NFs surface. Irregular small and well distributed particles along the NFs are shown in Figure 5A. Figure 5B shows the HR-TEM images, as can be seen, the formed NPs have different crystal lattice structure than the matrix CNFs. Furthermore, the NPs have crystalline structure; however, the matrix has the amorphous structure. Thus, one can claim that the NFs are carbon and the NPs are NiCu alloy. The fringe distance was found to be 1.83° A which is agreeable with the (200) plane of $Cu_{3.8}Ni$ alloy. Accordingly, the obtained nanofibers are CNF matrix decorated by NiCu nanoparticles. However, the metallic nanoparticles are sheathed inside a very thin carbon shell. Thin carbon layer may hinder metallic nanoparticle corrosion and may also prohibit agglomeration during chemical reactions.

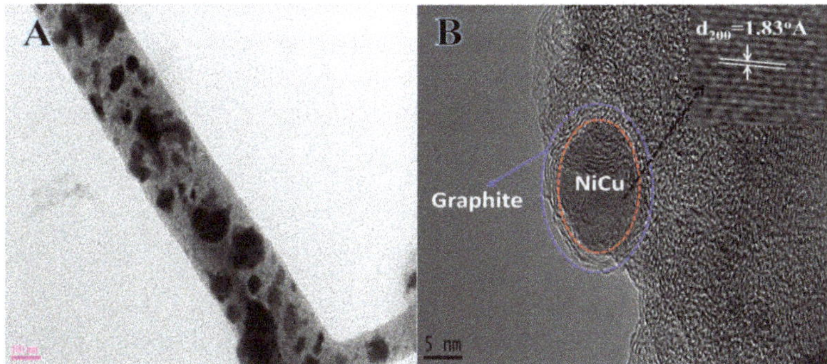

Figure 5. (**A**) TEM image of a single nanofiber and (**B**) high resolution TEM image.

To confirm aforementioned hypothesis and nicely understand the composition and elemental analysis of the fabricated NFs, STEM and TEM-EDX has been investigated. Figure 6A shows STEM image of a single NF. It is very clear from the image that irregular small metallic NPs decorated the CNFs. Figure 6B indicates the line EDX results corresponding to the line shown in Figure 6A. As seen in Figure 6B, copper and nickel have the identical elemental distribution along with the selected line in Figure 6A. It demonstrates good alloying features of the obtained metallic NPs which confirms the XRD result. Furthermore, the carbon elemental distribution proves the covering of the bimetallic NPs with thin layer of carbon. The catalytic activity of the fabricated NFs might be enhanced due to the occurrence of carbon which acts as an adsorbent in any catalytic chemical reactions by improving the attachment of the reactants with the catalytic material.

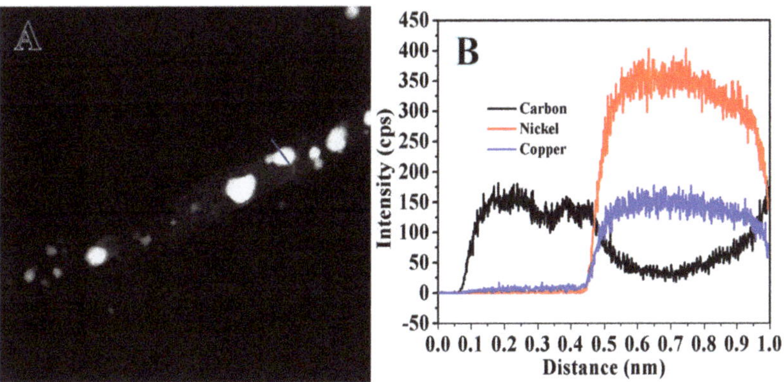

Figure 6. STEM image for a single nanofiber along with the line EDX analysis (**A**) and TEM EDX results for the line in A (**B**).

2.2. Electrochemical Analysis

Figure 7 presents cyclic voltammetry (CV) behaviors of the prepared sample in 1 M KOH after repeated cyclization for 30 cycles at 50 mV/s. A pair of redox peaks in the forward and backward scans is observed at potential values of 560 and 226 mV, respectively. These peaks can be attributed to the oxidation of Ni^{++} ($Ni(OH)_2$) to the active Ni^{+++} (NiOOH) layer [2,13]. Consecutive cyclic voltammograms show progressive CD increases for the cathodic peak as a result of the continuous entry of OH^- ions into the $Ni(OH)_2$ layer to form a thickened NiOOH layer from Ni^{++}/Ni^{+++} transformation.

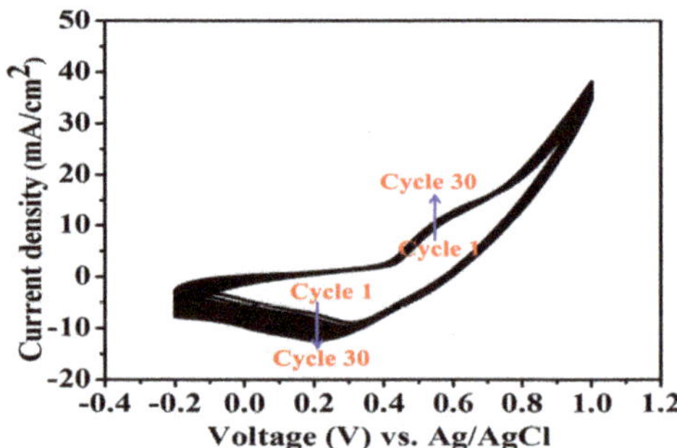

Figure 7. Consecutive CV of NiCu-CNF electrodes in 1 M KOH solution at 50 mV/s for 30 cycles.

The CV behavior of NiCu-CNF electrodes in 1 M KOH solution is compared before and after introducing urea molecules (1 M) at 50 mV/s (Figure 8A). NiOOH species act as the main electrocatalysts for urea molecule oxidation. Accordingly, oxidation peak formation observed in the presence of urea molecules begins with the development of NiOOH at the electrocatalyst surface at a potential value of 400 mV versus Ag/AgCl electrode, forming an oxidation peak with a potential value of 800 mV after the complete formation of all NiOOH species. The UO produces of 98 mA/cm² relative to that of 34 mA/cm² found in the absence of urea, indicating the enhanced performance of synthesized composite NFs in UO reactions. It is noteworthy to mention that the high current density obtained above 0.8 V may be attributed to oxygen evolution reaction (OER) in the alkaline media in the urea free solution. However, this phenomenon does not occur in the presence of organic molecules in the alkali solution since it needs potential more than 1 V to obtain OER [34]. Accordingly, the current produced in the presence of urea solution is regard to the UO [34,35]. Compared to urea-free solution, a decrease on the cathodic peak current with the addition of urea could be due to the further oxidation of urea and/or the intermediated compounds which consume the active layer and convert NiO(OH) to $Ni(OH)_2$ [35]. This conversion happened at a 226 mV in absence of urea while the little decrease in the potential in the presence of urea could be explained by the formation of nickel oxides of different morphologies. Abdel Kareem et al. [2] studied the electrocatalytic activity of Ni-Cd CNFs electrodes for oxidizing urea at different urea solution concentrations at 50 mV/s. A high CD of 68 mA/cm² at anodic peak potential value of 0.68 V and low onset potential of 0.35 V in (1 M urea + 1 M KOH) solution was measured as a higher CD and onset potential values by roughly 16 mA/cm² and 0.03 V, respectively, relative to that of our prepared composite NFs. Table 1 shows current densities obtained from recent studies. The introduced composite NFs exhibit an elevated CD and minimal onset potential (380 mV versus Ag/AgCl electrode). This can be attributed to the synergistic effects of copper and nickel. Furthermore, in alkaline media, copper is used to enhance electrocatalytic activity of nickel in alkaline media without participate in the redox reactions. In addition, it improves the electrocatalytic activity of Ni by filling Ni d-band vacancies with Cu electrons, which limit the volume expansion of the Ni^{2+} phase during oxidation [27,36]. In addition, thin carbon layers covered with bimetallic NiCu (Figure 5B) can promote the electron transfer process, enhance the chemical and corrosion resistance of bimetallic NiCu during UO and increase the adsorption of urea molecules. To confirm the aforementioned hypothesis, CV behavior of glassy carbon electrode, graphite, CuNPs, and NiCu-CNF electrodes in (1 M Urea + 1M KOH) solution at 50 mV/s (Figure 8B) have been achieved. As can be seen, the fabricated NFs have the high electrocatalytic activity towards UO compared to glassy carbon electrode, graphite, and Cu NPs.

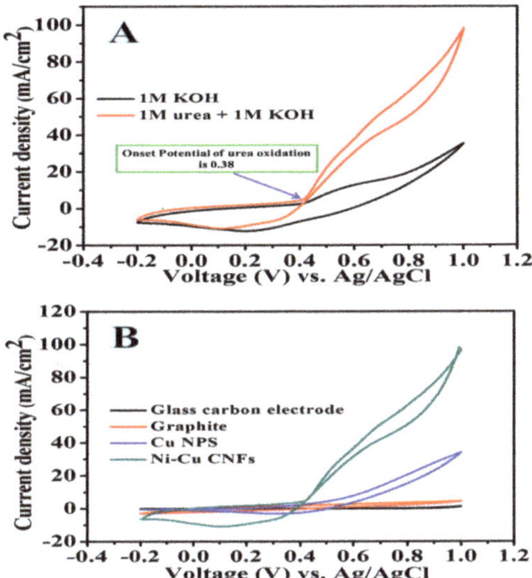

Figure 8. (**A**) CV of NiCu-CNF electrodes in 1 M KOH solution before and after adding 1 M urea solution at 50 mV/s and (**B**) study the influence of glass carbon electrode, graphite, Cu NPs, and Ni-Cu CNFs on urea electroxidation (1 M urea + 1 M KOH, scan rate of 50 mV/s).

Table 1. CD values of different electrocatalysts.

Electrocatalyst	Onset Potential	CD (mA/cm²)	Anodic Peak Potential (V)	Reference Electrode	Urea Concentration (M) in KOH (M)	Reference
Ni	0.39	22	0.5	Hg/HgO	0.33 in 1	[4]
Ni-Zn-Co	0.35	24	0.5	Hg/HgO	0.33 in 1	
Ni-P	1.37	~40	1.5	RHE	0.33 in 1	[37]
NiO/Gt	0.345	17	0.62	Ag/AgCl	0.33 in 0.5M NaOH	[38]
NiO/Gt-15	0.345	17	0.64	Ag/AgCl		[1]
$Ni_{1.5}Mn_{1.5}O_4$	0.29	6.9	0.5	Ag/AgCl	0.33 in 1	[39]
Ni-ERGO	0.4	35		Hg/HgO	0.33 in 1	[40]
$NiCo(OH)_2$	0.25	~20	~0.42	Hg/HgO	0.33 in 1	[18]
Ni-Cu/ZnO@MWCNT	~30	32	~0.45	Ag/AgCl	0.07 in 0.4	[41]
NiMn-CNFs	0.29	67	0.58	Ag/AgCl	2 in 1	[6]
Ni-Cu-CNFs	0.38	25	0.6	Ag/AgCl	1 in 1	This study

Figure 9A illustrates the dependence of the electrocatalytic activity of the prepared composite NFs electrocatalyst on urea concentrations at 50 mV/s. Increasing urea concentrations from 1/3 M to 1 M resulted in a higher CD. Thus, it can be concluded that diffusion governs reactions within this concentration domain. When urea concentrations increased to 2 M, the CD was not changed (data are not shown). This can be attributed to the fact that at high urea concentrations, the process no longer depends on diffusion and it is controlled by kinetics limitations. Thus, after a value of 1 M is reached, urea concentrations do not influence the CD. This result implies that the surface was covered by high concentrated urea molecules which limited hydroxyl groups (OH^-) required for oxidation of extra urea molecules [13].

The dependence of the logarithmic value of urea OCD on urea concentrations is shown in Figure 9B. A linear relationship was achieved. The slope of this linear relation is an estimate of the reaction order with respect to a urea concentration of 0.51 in agreement with the values obtained by others [21,42].

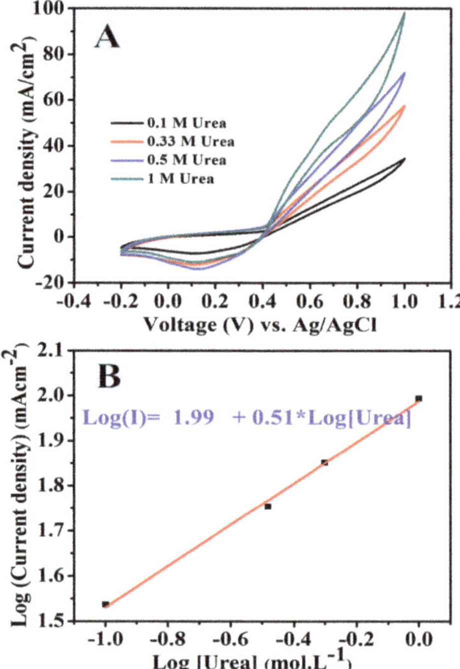

Figure 9. (**A**) Cyclic voltammograms of UO in composite NF anodes in 1 M KOH solution for 50 mV/s containing different concentrations of urea. (**B**) Double logarithmic variations of CD of UO with urea concentrations.

The influence of scan rates (10, 30, 50, 100, 200, and 300 mV/s) on the electrocatalytic activity of $Cu_{3.8}Ni$ NPs-anchored carbon electrocatalyst for UO in (1 M urea + 1 M KOH) solution is studied as shown in Figure 10A. Anodic and cathodic currents were enhanced with increasing scan rates. A linear relationship is obtained by plotting the square root of the scan rate as a function of the CD of the UO peak in Figure 10B. UO reaction is predicted to be a diffusion-governing operation. Furthermore, the variation in the scan rate is normalized current with the scan rate in Figure 10C shows a characteristic curve for an irreversible oxidation process. Accordingly, the presence of EC (electrochemical-chemical) mechanisms is proposed [42]. The diffusion coefficient of urea molecules at the electrocatalyst surface can be determined from the following equation [6,43]:

$$I_p = 0.4958 \times 10^{-3} \, nFC_{urea} \, (\alpha nF/RT)^{0.5} \, D^{0.5} \, v^{0.5} \tag{4}$$

D is the diffusion coefficient of urea, v is the scan rate, α is the anodic transfer coefficient, n is the number of transferred electrons, C_{urea} is urea concentrations, and F is Faraday's constant. According to the literature survey, n (2.8) was used to denote ethanol, which is more similar in chemical composition to urea than methanol and α value is 0.1 [6,43]. From a linear analysis of the relationship between I_p at the anodic peak and $v^{0.5}$, the diffusion coefficient is calculated as $6.04 \times 10^{-3} \, cm^2/s$.

The long-term stability of our fabricated NFs was examined via chronoamperometry with a potential value of 600 mV (Ag/AgCl) for 900 s as shown in Figure 10D. The introduced electrocatalyst underwent current decay and then reached a steady state. This decrease in the current can be attributed to the consumption of urea from solution. Additionally, the performance of electrodes was not affected due to the use of introduced catalysts supporting their enhanced stability.

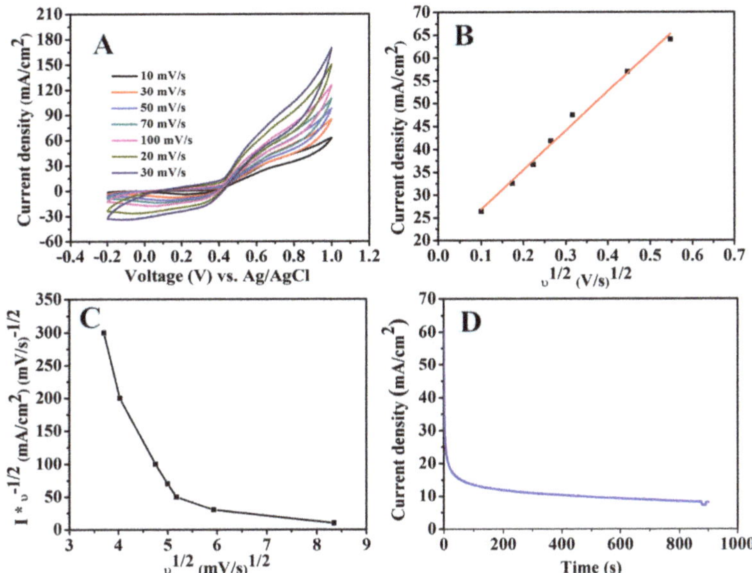

Figure 10. (**A**) CV of composite NF electrodes in 1 M KOH aqueous solution after adding 1 M of urea at different scan rate values (10, 30, 50, 70, 100, 200, and 300 mV/s). (**B**) Variations in the square root of the scan rate as a function of the CD of UO at composite NF electrodes. (**C**) The scan rate normalized current density plotted against the scan rate. (**D**) Chronoamperogram of composite NF electrodes in 1 M KOH aqueous solution containing 1 M of urea solution at 600 mV (Ag/AgCl) for 900 s.

3. Materials and Methods

Poly(vinyl alcohol) (PVA) (*MW* = 85000 − 124000 g/mol) was purchased from DC Chemical Co., Seoul, South Korea. Copper(II) acetate monohydrate (CuAc·H_2O, 98%) and nickel(II) acetate tetrahydrate (NiAc·$4H_2O$, 98%) were purchased from Sigma Aldrich, Seoul, South Korea. Deionized water (DI) was used as a solvent.

To fabricate electrospun NFs, PVA electrospinning polymeric solution was constructed by dissolving 10 wt% of PVA into DI water. The solution was then stirred until a clear solution was obtained. Then 0.2 g of copper(II) acetate monohydrate and 0.8 g of nickel(II) acetate tetrahydrate were dissolved in 5 g of DI water. The prepared metal precursor solution was then added to a glass bottle with 15 g of PVA polymeric solution. The mixture was stirred for 6 h at 60 °C and then cooled at room temperature.

The prepared sol-gel solution was electrospun using a lab scale electrospinning machine at a high voltage of 18 kV using a DC power supply and maintaining a distance of 18 cm. Fabricated NFs were gathered on a rotating cylinder covered with aluminum foil. The prepared NFs samples were dried under a vacuum for 1 day at 50 °C. Finally, the products were calcined under an argon atmosphere (Ar) for 5 h at 800 °C (2.3 °C/min to obtain final product).

A JEOL JSM-5900 scanning electron microscope (JEOL Ltd., Tokyo, Japan) and field-emission scanning electron microscope (FESEM, Hitachi S-7400, Tokyo, Japan) were used to study the surface morphology of the formed NFs. A Rigaku X-ray diffractometer (Rigaku Co., Tokyo, Japan) with Cu Kα (λ = 1.54056 Å) radiation over a range of 2θ angles of 10° to 80° was utilized to investigate the phase and crystallinity of the fabricated NFs. To acquire high-resolution images, the NFs were characterized using a JEOL JEM-2200FS transmission electron microscope (TEM) operated at 200 kV and equipped with EDX (JEOL Ltd., Tokyo, Japan). A thermo gravimetric analyzer (TGA, Ptris1, Perkin Elmer,

Billerica, Massachusetts, USA) was used to study the thermal stability of NFs by heating the samples to 750 °C under an oxygen atmosphere.

Cyclic voltammetry and chronoamperometry experiments were conducted using autolab potentiostate ((PGSTAT101 controlled by Nova software), Metrohm Autolab Co., Kanaalweg 29-G, 3526 KM, Utrecht, Netherlands). The system includes three electrodes: a counter electrode (platinum wire), reference electrode (Ag/AgCl, saturated KCl) and working electrode (fabricated electrospun NFs). Measurements were performed using different concentrations of alkaline urea solution and the range of the sweep potential of the electrochemical cell was set from to −0.2 to 1 V (Ag/AgCl).

The working electrode was prepared by creating a slurry with 2 mg of the fabricated NFs, 20 μL of Nafion solution (5 wt%) and 400 μL of isopropanol. The slurry was then sonicated at room temperature for 30 min. Finally, 15 μL of the prepared slurry was poured onto the active area of the glassy carbon electrode, which was then dried at 80 °C for 20 min. The utilized active area of the glassy carbon electrode (the electrode has circle shape with D = 3 mm) was 0.0713 cm^2. Therefore, all the obtained data was normalized to aforementioned area.

4. Conclusions

NiCu NP-decorated carbon NFs were synthesized using an electrospinning machine. Composite NFs exhibited strong catalytic activity for urea electro-oxidation as inferred from increased OCD (98 mA cm^{-2}) and lowered onset potential values (380 mV versus (Ag/AgCl)). The reaction order of composite NF electrodes with respect to urea solution concentrations was found to be 0.51. Urea electro-oxidation reactions observed at the studied composite NFs were irreversible.

Funding: This research was funded by SABIC Company and Jazan University, grant number (Sabic 3/2018/1).

Acknowledgments: The author would like to thank SABIC Company and Jazan University for financially supporting this project.

Conflicts of Interest: The author declares no conflict of interest.

References

1. Abdel Hameed, R.M.; Medany, S.S. Enhanced electrocatalytic activity of NiO nanoparticles supported on carbon planes towards urea electro-oxidation in NaOH solution. *Int. J. Hydrogen Energy* **2017**, *42*, 24117–24130. [CrossRef]
2. Abdelkareem, M.A.; Al Haj, Y.; Alajami, M.; Alawadhi, H.; Barakat, N.A. Ni-Cd carbon nanofibers as an effective catalyst for urea fuel cell. *J. Environ.. Chem. Eng.* **2018**, *6*, 332–337. [CrossRef]
3. Wang, L.; Du, T.; Cheng, J.; Xie, X.; Yang, B.; Li, M. Enhanced activity of urea electrooxidation on nickel catalysts supported on tungsten carbides/carbon nanotubes. *J. Power Sources* **2015**, *280*, 550–554. [CrossRef]
4. Yan, W.; Wang, D.; Botte, G.G. Electrochemical decomposition of urea with Ni-based catalysts. *Appl. Catal. B* **2012**, *127*, 221–226. [CrossRef]
5. Yan, W.; Wang, D.; Diaz, L.A.; Botte, G.G. Nickel nanowires as effective catalysts for urea electro-oxidation. *Electrochim. Acta* **2014**, *134*, 266–271. [CrossRef]
6. Barakat, N.A.; El-Newehy, M.H.; Yasin, A.S.; Ghouri, Z.K.; Al-Deyab, S.S. Ni&Mn nanoparticles-decorated carbon nanofibers as effective electrocatalyst for urea oxidation. *Appl. Catal. A* **2016**, *510*, 180–188.
7. Boggs, B.K.; King, R.L.; Botte, G.G. Urea electrolysis: direct hydrogen production from urine. *Chem. Commun.* **2009**, 4859–4861. [CrossRef] [PubMed]
8. Ye, K.; Zhang, D.; Guo, F.; Cheng, K.; Wang, G.; Cao, D. Highly porous nickel@ carbon sponge as a novel type of three-dimensional anode with low cost for high catalytic performance of urea electro-oxidation in alkaline medium. *J. Power Sources* **2015**, *283*, 408–415. [CrossRef]
9. Wang, D.; Yan, W.; Vijapur, S.H.; Botte, G.G. Enhanced electrocatalytic oxidation of urea based on nickel hydroxide nanoribbons. *J. Power Sources* **2012**, *217*, 498–502. [CrossRef]
10. Bian, L.; Du, Q.; Luo, M.; Qu, L.; Li, M. Monodisperse nickel nanoparticles supported on multi-walls carbon nanotubes as an effective catalyst for the electro-oxidation of urea. *Int. J. Hydrogen Energy* **2017**, *42*, 25244–25250. [CrossRef]

11. Alajami, M.; Yassin, M.A.; Ghouri, Z.K.; Al-Meer, S.; Barakat, N.A. Influence of bimetallic nanoparticles composition and synthesis temperature on the electrocatalytic activity of NiMn-incorporated carbon nanofibers toward urea oxidation. *Int. J. Hydrogen Energy* **2018**, *43*, 5561–5575. [CrossRef]

12. Barakat, N.A.; Alajami, M.; Al Haj, Y.; Obaid, M.; Al-Meer, S. Enhanced onset potential NiMn-decorated activated carbon as effective and applicable anode in urea fuel cells. *Catal. Commun.* **2017**, *97*, 32–36. [CrossRef]

13. Barakat, N.A.M.; Motlak, M.; Ghouri, Z.K.; Yasin, A.S.; El-Newehy, M.H.; Al-Deyab, S.S. Nickel nanoparticles-decorated graphene as highly effective and stable electrocatalyst for urea electrooxidation. *J. Mol. Catal. A Chem.* **2016**, *421*, 83–91. [CrossRef]

14. Guo, F.; Ye, K.; Cheng, K.; Wang, G.; Cao, D. Preparation of nickel nanowire arrays electrode for urea electro-oxidation in alkaline medium. *J. Power Sources* **2015**, *278*, 562–568. [CrossRef]

15. Barakat, N.A.M.; Amen, M.T.; Al-Mubaddel, F.S.; Karim, M.R.; Alrashed, M. NiSn nanoparticle-incorporated carbon nanofibers as efficient electrocatalysts for urea oxidation and working anodes in direct urea fuel cells. *J. Adv. Res.* **2019**, *16*, 43–53. [CrossRef]

16. Hameed, R.A.; Tammam, R.H. Nickel oxide nanoparticles grown on mesoporous carbon as an efficient electrocatalyst for urea electro-oxidation. *Int. J. Hydrogen Energy* **2018**, *43*, 20591–20606. [CrossRef]

17. Wang, G.; Ye, K.; Shao, J.; Zhang, Y.; Zhu, K.; Cheng, K.; Yan, J.; Wang, G.; Cao, D. Porous Ni2P nanoflower supported on nickel foam as an efficient three-dimensional electrode for urea electro-oxidation in alkaline medium. *Int. J. Hydrogen Energy* **2018**, *43*, 9316–9325. [CrossRef]

18. Yan, W.; Wang, D.; Botte, G.G. Nickel and cobalt bimetallic hydroxide catalysts for urea electro-oxidation. *Electrochim. Acta* **2012**, *61*, 25–30. [CrossRef]

19. Vedharathinam, V.; Botte, G.G. Direct evidence of the mechanism for the electro-oxidation of urea on Ni (OH) 2 catalyst in alkaline medium. *Electrochim. Acta* **2013**, *108*, 660–665. [CrossRef]

20. Abutaleb, A.; Lolla, D.; Aljuhani, A.; Shin, H.U.; Rajala, J.W.; Chase, G.G. Effects of Surfactants on the Morphology and Properties of Electrospun Polyetherimide Fibers. *Fibers* **2017**, *5*, 33. [CrossRef]

21. Al-Enizi, A.M.; Brooks, R.M.; El-Halwany, M.; Yousef, A.; Nafady, A.; Hameed, R.A. CoCr7C3-like nanorods embedded on carbon nanofibers as effective electrocatalyst for methanol electro-oxidation. *Int. J. Hydrogen Energy* **2018**, *43*, 9943–9953. [CrossRef]

22. Lolla, D.; Lolla, M.; Abutaleb, A.; Shin, H.U.; Reneker, D.H.; Chase, G.G. Fabrication, polarization of electrospun polyvinylidene fluoride electret fibers and effect on capturing nanoscale solid aerosols. *Materials* **2016**, *9*, 671. [CrossRef]

23. Shin, H.U.; Abutaleb, A.; Lolla, D.; Chase, G.G. Effect of Calcination Temperature on NO–CO Decomposition by Pd Catalyst Nanoparticles Supported on Alumina Nanofibers. *Fibers* **2017**, *5*, 22. [CrossRef]

24. Yousef, A.; Akhtar, M.S.; Barakat, N.A.M.; Motlak, M.; Yang, O.B.; Kim, H.Y. Effective NiCu NPs-doped carbon nanofibers as counter electrodes for dye-sensitized solar cells. *Electrochim. Acta* **2013**, *102*, 142–148. [CrossRef]

25. Yousef, A.; Barakat, N.A.M.; El-Newehy, M.; Kim, H.Y. Chemically stable electrospun NiCu nanorods@ carbon nanofibers for highly efficient dehydrogenation of ammonia borane. *Int. J. Hydrogen Energy* **2012**, *37*, 17715–17723. [CrossRef]

26. Yousef, A.; Brooks, R.M.; Abdelkareem, M.A.; Khamaj, J.A.; El-Halwany, M.; Barakat, N.A.; EL-Newehy, M.H.; Kim, H.Y. Electrospun NiCu Nanoalloy Decorated on Carbon Nanofibers as Chemical Stable Electrocatalyst for Methanol Oxidation. *ECS Electrochem. Lett.* **2015**, *4*, F51–F55. [CrossRef]

27. Yousef, A.; Brooks, R.M.; El-Halwany, M.M.; Abdelkareem, M.A.; Khamaj, J.A.; EL-Newehy, M.H.; Barakat, N.A.M.; Kim, H.Y. Fabrication of Electrical Conductive NiCu– Carbon Nanocomposite for Direct Ethanol Fuel Cells. *Int. J. Electrochem. Sci.* **2015**, *10*, 7025–7032.

28. Yousef, A.; Brooks, R.M.; Abutaleb, A.; El-Newehy, M.H.; Al-Deyab, S.S.; Kim, H.Y. One-step synthesis of Co-TiC-carbon composite nanofibers at low temperature. *Ceram. Int.* **2017**, *43*, 15735–15742. [CrossRef]

29. Yousef, A.; Brooks, R.M.; El-Halwany, M.M.; Abutaleb, A.; El-Newehy, M.H.; Al-Deyab, S.S.; Kim, H.Y. Electrospun CoCr7C3-supported C nanofibers: Effective, durable, and chemically stable catalyst for H2 gas generation from ammonia borane. *Mol. Catal.* **2017**, *434*, 32–38. [CrossRef]

30. Yousef, A.; Brooks, R.M.; El-Newehy, M.H.; Al-Deyab, S.S.; Kim, H.Y. Electrospun Co-TiC nanoparticles embedded on carbon nanofibers: Active and chemically stable counter electrode for methanol fuel cells and dye-sensitized solar cells. *Int. J. Hydrogen Energy* **2017**, *42*, 10407–10415. [CrossRef]

31. Barakat, N.A.; Abdelkareem, M.A.; El-Newehy, M.; Kim, H.Y. Influence of the nanofibrous morphology on the catalytic activity of NiO nanostructures: an effective impact toward methanol electrooxidation. *Nanoscale Res. Lett.* **2013**, *8*, 402. [CrossRef] [PubMed]

32. Boskovic, B.O.; Stolojan, V.; Zeze, D.A.; Forrest, R.D.; Silva, S.R.P.; Haq, S. Branched carbon nanofiber network synthesis at room temperature using radio frequency supported microwave plasmas. *J. Appl. Phys.* **2004**, *96*, 3443. [CrossRef]

33. Ud Din, I.; Shaharun, M.S.; Subbarao, D.; Naeem, A. Surface modification of carbon nanofibers by HNO3 treatment. *Ceram. Int.* **2016**, *42*, 966–970. [CrossRef]

34. Rahim, M.A.; Hameed, R.A.; Khalil, M. Nickel as a catalyst for the electro-oxidation of methanol in alkaline medium. *J. Power Sources* **2004**, *134*, 160–169. [CrossRef]

35. Thamer, B.M.; El-Newehy, M.H.; Barakat, N.A.M.; Abdelkareem, M.A.; Al-Deyab, S.S.; Kim, H.Y. Influence of Nitrogen doping on the Catalytic Activity of Ni-incorporated Carbon Nanofibers for Alkaline Direct Methanol Fuel Cells. *Electrochim. Acta* **2014**, *142*, 228–239. [CrossRef]

36. Jafarian, M.; Moghaddam, R.; Mahjani, M.; Gobal, F. Electro-Catalytic Oxidation of Methanol on a Ni–Cu Alloy in Alkaline Medium. *J. Appl. Electrochem.* **2006**, *36*, 913–918. [CrossRef]

37. Ding, R.; Li, X.; Shi, W.; Xu, Q.; Wang, L.; Jiang, H.; Yang, Z.; Liu, E. Mesoporous Ni-P nanocatalysts for alkaline urea elect oxidation. *Electrochim. Acta* **2016**, *222*, 455–462. [CrossRef]

38. Abdel Hameed, R.M.; Medany, S.S. Influence of support material on the electrocatalytic activity of nickel oxide nanoparticles for urea electro-oxidation reaction. *J. Colloid Interface Sci.* **2018**, *513*, 536–548. [CrossRef]

39. Periyasamy, S.; Subramanian, P.; Levi, E.; Aurbach, D.; Gedanken, A.; Schechter, A. Exceptionally Active and Stable Spinel Nickel Manganese Oxide Electrocatalysts for Urea Oxidation Reaction. *Appl. Mater. Interfaces* **2016**, *8*, 12176–12185. [CrossRef]

40. Wang, D.; Yan, W.; Vijapur, S.H.; Botte, G.G. Electrochimica Acta Electrochemically reduced graphene oxide–nickel nanocomposites for urea electrolysis. *Electrochim. Acta* **2013**, *89*, 732–736. [CrossRef]

41. Basumatary, P.; Konwar, D.; Yoon, Y.S. A novel NieCu/ZnO@MWCNT anode employed in urea fuel cell to attain superior performances. *Electrochim. Acta* **2018**, *261*, 78–85. [CrossRef]

42. Yousef, A.; El-Newehy, M.H.; Al-Deyab, S.S.; Barakat, N.A. Facile synthesis of Ni-decorated multi-layers graphene sheets as effective anode for direct urea fuel cells. *Arabian J. Chem.* **2017**, *10*, 811–822. [CrossRef]

43. Liu, J.; Ye, J.; Xu, C.; Jiang, S.P.; Tong, Y. Kinetics of ethanol electrooxidation at Pd electrodeposited on Ti. *Electrochem. Commun.* **2007**, *9*, 2334–2339. [CrossRef]

 *catalysts*

Article

Influence of Sn Content, Nanostructural Morphology, and Synthesis Temperature on the Electrochemical Active Area of Ni-Sn/C Nanocomposite: Verification of Methanol and Urea Electrooxidation

Nasser A. M. Barakat [1,*], Mohammad Ali Abdelkareem [1,2,3] and Emad A. M. Abdelghani [1]

[1] Chemical Engineering Department, Minia University, El-Minia 61519, Egypt; mabdulkareem@sharjah.ac.ae (M.A.A.); omaremad@yahoo.com (E.A.M.A.)

[2] Dept. of Sustainable and Renewable Energy Engineering, University of Sharjah, P.O. Box 27272 Sharjah, UAE

[3] Center for Advanced Materials Research, University of Sharjah, P.O. Box 27272 Sharjah, UAE

* Correspondence: nasbarakat@minia.edu.eg; Tel.: +20-8623234008

Received: 12 February 2019; Accepted: 22 March 2019; Published: 3 April 2019

Abstract: In contrast to precious metals (e.g., Pt), which possess their electro catalytic activities due to their surface electronic structure, the activity of the Ni-based electrocatalysts depends on formation of an electroactive surface area (ESA) from the oxyhydroxide layer (NiOOH). In this study, the influences of Sn content, nanostructural morphology, and synthesis temperature on the ESA of Sn-incorporated Ni/C nanostructures were studied. To investigate the effect of the nanostructural, Sn-incorporated Ni/C nanostructures, nanofibers were synthesized by electrospinning a tin chloride/nickel acetate/poly (vinyl alcohol) solution, followed by calcination under inert atmosphere at high temperatures (700, 850, and 1000 °C). On the other hand, the same composite was formulated in nanoparticulate form by a sol-gel procedure. The electrochemical measurements indicated that the nanofibrous morphology strongly enhanced formation of the ESA. Investigation of the tin content concluded that the optimum co-catalyst content depends on the synthesis temperature. Typically, the maximum ESA was observed at 10 and 15 wt % of the co-catalyst for the nanofibers prepared at 700 and 850 °C, respectively. Study of the effect of synthesis temperature concluded that at the same tin content, 850 °C calcination temperature reveals the best activity compared to 700 and 1000 °C. Practical verification was achieved by investigation of the electrocatalytic activity toward methanol and urea oxidation. The results confirmed that the activity is directly proportionate to the ESA, especially in the case of urea oxidation. Moreover, beside the distinct increase in the current density, at the optimum calcination temperature and co-catalyst content, a distinguished decrease in the onset potential of both urea and methanol oxidation was observed.

Keywords: electroactive surface area; electrospinning; Sn-incorporated Ni/C nanofibers; Methanol; Urea

1. Introduction

The near depletion of fossil fuels, as well as emission of various environmentally unacceptable oxides, such as NO_x, CO_x, and SO_x, has forced researchers at both academic and industrial levels to search for alternative fuel sources. Fuel cells could be promising candidates if the capital cost is reduced. Commercially, utilizing precious electrodes strongly constrains the wide applications of the fuel cells because of the high cost. Although in literature there are huge trials investigating the use of the transition metals as electrodes, based on our best knowledge, there are no commercial applications. On the other hand, transition metal electrodes can be effectively used as anode materials in hydrogen production by electrolysis. Hydrogen is the most promising candidate fuel from sustainability and environmental points of view, and as a future alternative to fossil fuels [1,2]. There are several techniques that have been introduced to generate hydrogen, including fermentation of waste biomasses [3,4], water electrolysis [5], water photosplitting [6], thermal production [7], etc. These general techniques suffer from some drawbacks, such as low yield (fermentation and water photosplitting), high cost (water electrolysis and thermal processes), or seasonality (water photosplitting under solar radiation). Among these methods, electrolysis is the most widely used due to the ability to reduce the required power by using other precursors rather than water, such as alcohols and urea [8–10].

Methanol is the simplest and the most popular investigated alcohol in the electrooxidation process. Methanol can be classified as a renewable energy source because it can be produced from biomasses [11,12]. Furthermore, compared to compressed hydrogen, methanol exhibits more efficient storage in terms of weight and volume. Accordingly, methanol does not require a cryogenic container that oftentimes needs to be maintained at a very low temperature [13,14]. Additionally, the risk of flash fire and explosion is low due to the low volatility. Therefore, hydrogen extraction from methanol fulfills most of the requirements for the generation and storing processes. Compared to water electrolysis, methanol electrooxidation needs a smaller potential at 1.2 and 0.02 V, respectively, which makes it a preferable technology from an energy requirement point of view. Methanol electrooxidation for hydrogen production can be performed according to the following reactions:

In acid media

$$\text{Anode:} \qquad CH_3OH + H_2O \rightarrow CO_2 + 6H^+ + 6e \tag{1}$$

$$\text{Cathode:} \qquad 6H^+ + 6e \rightarrow 3H_2 \tag{2}$$

$$\text{Overall:} \qquad CH_3OH + H_2O \rightarrow CO_2 + 3H_2 \tag{3}$$

In alkaline media

$$\text{Anode:} \qquad CH_3OH + 6OH^- \rightarrow CO_2 + 5H_2O + 6e \tag{4}$$

$$\text{Cathode:} \qquad 6H_2O + 6e \rightarrow 3H_2 + 6OH^- \tag{5}$$

$$\text{Overall:} \qquad CH_3OH + H_2O \rightarrow CO_2 + 3H_2 \tag{6}$$

Yet, this trivial required cell potential is a theoretical value, while a high overpotential is created depending on the electrocatalytic activity of the utilized anode [15,16].

Beside methanol, urea is another important future candidate as a hydrogen precursor because urea-contaminated wastewater is produced in large amounts from fertilizer manufacturing plants. Therefore, generating hydrogen from urea-polluted wastewater solves an environmental problem simultaneously. Theoretically, production of hydrogen from urea does not need additional power, as it is an exothermic reaction according to the following equations [17–20]:

$$\text{Anode:} \qquad CO(NH_2)_2 + 6OH^- \rightarrow N_2 + 5H_2O + CO_2 + 6e \qquad E^0 = -0.746 \text{ V} \tag{7}$$

$$\text{Cathode:} \qquad 6H_2O + 6e \rightarrow 3H_2 + 6OH^- \qquad E^0 = -0.829 \text{ V} \tag{8}$$

$$\text{Overall:} \qquad CO(NH_2)_2 + H_2O \rightarrow N_2 + 3H_2 + CO_2 \qquad E^0 = -0.083 \text{ V} \tag{9}$$

So far, based on our best knowledge, no anode material has been reported that could achieve the task without additional power.

Among the investigated transition metals, Ni showed promising electrochemical oxidation activity for both methanol and urea [21,22]. In contrast to precious metals, nickel owes its electocatalytic activity from the ability to form an electroactive oxyhydroxide layer (NiOOH), which can be formed by being swept in an alkaline solution or during the oxidation process [23]. The active compound forms an electroactive surface area (ESA). Typically, nickel is first oxidized to $Ni(OH)_2$, then the hydroxide layer is transformed to the oxy(hydroxide) form [24,25]. Therefore, there is a strong belief that enhancing the formation of the ESA leads to improvement of the electrocatalytic performance.

Different strategies have been carried out to improve activity of Ni, whether by increasing its surface area [26–29] or by forming an alloy with other transition metals, such as Mn [30], Co [31,32], Cr [33], Cu [34], and TiO_2 [35]. As a co-catalyst, tin is extensively used to improve the electrochemical oxidation activity of Pt towards methanol oxidation [36–40]. Tin improves methanol oxidation by one or more of the following ways. By providing a charge to Pt, it can increase the adsorption of alcoholic residues on the catalyst surface, and it can decrease the content of higher valent Pt sites [37]. Moreover, tin can decrease the poisoning of Pt by CO by providing an oxygen group on its surface [40]. With nickel, NiSn bimetallic nanoparticle-immobilized titanium and carbon nanofiber supports were introduced as electrode materials for methanol and urea electrooxidation [20,41,42]. Moreover, NiSn discrete nanoparticles have been reported as a stable electro catalyst for methanol oxidation [43].

Beside the catalyst composition, the size also has a strong influence on the activity. Accordingly, synthesizing the functional materials in nanostructural forms strongly enhances the performance. However, the nanostructural morphology can considerably affect the activity. The large axial ratio characteristic improves the electron transfer process. This is a very effective feature that privileges the nanofibers from the other nanostructures [44]. There are several nanofibers making techniques that have been reported, including template-assisted [45], self-assembly [46], drawing [47], and electrospinning [48]. However, only the electrospinning process is widely used due to its simplicity, high yield, low cost, and ease of control on the product morphology [28,48,49]. Basically, the electrooxidation reactions are considered a combination between adsorption and chemical reaction processes. Therefore, the carbonaceous materials showed good performances as supports for different electrocatalysts [50–53]

In this study, the influence of tin content, the nanostructural morphology, and the synthesis temperature on the ESA of Sn-incorporated Ni/C nanostructures were investigated. To properly prove the obtained conclusion, practical verification was done by investigating the electocatalytic activities toward methanol and urea. The nanofibers have been prepared by calcination of electrospun nanofibers composed of poly(vinyl alcohol), tin chloride, and nickel acetate under vacuum atmosphere at different temperatures. Nanoparticulate morphology was formulated by crushing and grinding dried sol-gels from the same mixture followed by a calcination process. The electrochemical measurements indicated that the ESA can be improved by optimizing the Sn content and the calcination temperature. Moreover, the nanofibrous morphology leads to formation of high ESA compared to the nanoparticles. Regarding methanol and urea electrooxidation, the results proved that the electrocatalytic activity mainly depends on the ESA.

2. Results Discussion

2.1. Crystalline Structure, Surface Morphology, and Composition of the Prepared Materials

Among the nanofiber making techniques, electrospinning produces the highest quality from the morphology point of view. Usually, if the electrospinning parameters are optimized, good morphology can be obtained. On the other hand, maintaining the nanofibrous morphology during the calcination process depends on the utilized precursor for the inorganic nanofibers. Polycondensation is the main characteristic of a good precursor. In this regard, alkoxides are the best precursors. Due to the containing oxy-organic anion, metals acetates show good tendency for polycondensation [54]. Consequently, after the calcination process, good morphology nanofibers were obtained, as seen in Figure 1, which displays Scanning Electron Microscope (SEM) and Field Emission Scanning Electron Microscope (FE SEM) images for the powder obtained from 10 wt % tin chloride. It is noteworthy mentioning that all the formulations (at different Sn contents) reveal good morphology nanofibers with the appearance of some nanoparticles at high tin contents. Data are not shown.

Figure 1. SEM; (**A**) and FE-SEM; (**B**) images for the prepared nanofibers containing 10 wt % Sn, and calcined at 850 °C.

XRD analysis is a high confidence technique for investigating the crystal structure and composition of the crystalline materials. Tin and nickel can form a bimetallic alloy with a wide composition range [55]. Figure 2 displays the XRD pattern of the nanofibers obtained from an initial electrospun solution containing different concentrations from the tin precursor. As shown, change in the initial electrospun solution concentration resulted in variation in the final product composition. However, it can be concluded that the produced nanofibers are composed of two main metalic compunds; Ni_3Sn_2 and Ni_3Sn. The diffraction peaks for the crystal plans of (1 0 1), (0 0 2), (1 0 2), (1 1 0), (2 0 1), (1 1 2), (1 0 3), (2 0 2), and (2 1 1) at 2θ values of 30.8°, 34.75°, 43.55°, 44.65°, 55.09°, 57.58°, 59.85°, 63.91°, and 73.42° conclude the formation of Ni_3Sn_2 alloy according to the standard JCDPS database (#06-0414). On the other hand, the other diffraction peaks at 2θ values of 28.6°, 39.3°, 42.5°, 44.8°, and 59.3°, corresponding to the (1 0 1), (2 0 0), (0 0 2), (2 0 1), and (2 0 2) crystal planes, respectively, conclude the formation of Ni_3Sn alloy (JCDPS# 35-1362). It is noteworthy mentioning that the nanomorphology did not affect the XRD result. In other words, at the same composition, similar XRD patterns were obtained for the nanofiber and nanoparticle samples. Furthermore, change in the calcination temperature resulted in change in the intensity of the observed peaks in the XRD pattern (data are not shown), however the data concluded that all the prepared samples are composed of NiSn alloys. The observed peak at 2θ~25° referring to an experimental d spacing of 3.37 Å indicates formation of graphite-like carbon (d (002), JCPDS; 41-1487).

Figure 3A illustrates the internal structural of produced nanofibers, which was investigated by TEM analysis. As shown in the figure, the prepared nanofibers are composed of crystalline nanoparticles incorporated in an amorphous nanofibrous matrix. Based on the XRD results, the crystalline nanoparticles can be assigned to NiSn, while the nanofiber body consists of the formed amorphous graphite. Accordingly, it can be concluded that the final products are NiSn nanoparticle-incorporated carbon nanofibers. Figure 3B demonstrates the line elemental analysis at a randomly selected line. As shown, the result is in consistent with the XRD data. Obtaining similar distribution for the nickel and tin concludes the formation of the NiSn alloy, which supports the XRD results.

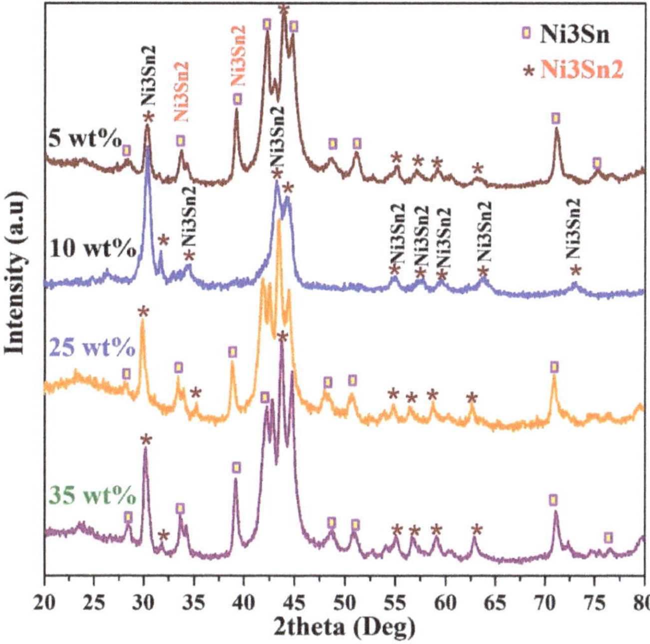

Figure 2. XRD analysis of the Sn-incorporated Ni/C nanofibers synthesized at 850 °C calcination temperature.

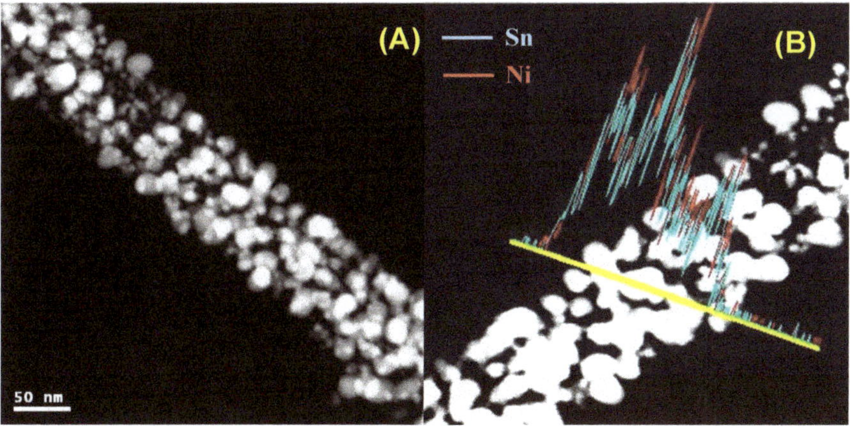

Figure 3. TEM image (**A**), and EDX line scan (**B**) for the prepared Sn-incorporated nanofibers; 10 wt % and 850 °C calcination temperature.

Change in the chemical composition can be explored by the thermal gravimetrical analysis (TGA); Figure 4A shows the obtained results. To mathematically extract the embedded information in the TGA spectrum, the first derivative of the graph was determined (Figure 4B) because the big changes in the sample weight are translated into peaks. The first peak, at ~100 °C, can be assigned to evaporation of the physically attached water in the sample. The high intensity peak observed at ~250 °C denotes a big decrease in the sample weight at this temperature. Based on the thermal characteristics of the used polymer, this peak can be assigned to the polymer decomposition [8,10]. Formation of zero valent metals rather than the expected metal oxides was attributed to the abnormal decomposition of the acetate anion under the inert atmosphere. Briefly, strongly reducing gases (namely carbon monoxide and hydrogen) are produced due to the decomposition of acetate, which leads to complete reduction of the precursors to metals. The full reduction of nickel acetate can be chemically represented by these equations [56]:

$$Ni(CH_3COO)_2 \cdot 4H_2O \rightarrow 0.86Ni(CH_3COO)_2 \cdot 0.14Ni(OH)_2 + 0.28CH_3COOH + 3.72H_2O \tag{10}$$

$$0.86\,Ni(CH3COO)_2 \cdot 0.14Ni(OH)_2 \rightarrow NiCO_3 + NiO + CH_3COCH_3 + H_2O \tag{11}$$

$$NiCO_3 \rightarrow 6NiO + CO_2 \tag{12}$$

$$NiO + CO \rightarrow 6Ni + CO_2 \tag{13}$$

Similarly, formation of pure tin can be assigned to the evolved reducing gases. Therefore, the remaining peaks in the first derivative plot represent the aforementioned chemical transformations.

As they were detected in the final product, it can be claimed that NiSn metallic alloys have higher melting points with respect to the synthesis temperature. On the other hand, tin (M.P 231.9 °C) is more volatile compared to nickel (M.P 1455 °C). Accordingly, vaporization of the free tin is expected during the calcination process, while, due to its high thermal stability, nickel can withstand at this high temperature without losing any weight. Therefore, free Ni (excess than the stoichiometry) can be expected to be present in the produced nanofibers in either amorphous form or embedded as discrete atoms inside the formed graphite; both structures are not detectable by the XRD analysis. Accordingly, it can be claimed that the nickel content in the final nanofibers matches the amount in the initial precursor.

$$Ni_{pre} = Ni_{Ni3Sn} + Ni_{Ni3Sn2} + Ni_{free} \tag{14}$$

On the other hand, no free tin can withstand the produced nanofibers, so the used tin can be estimated as follows:

$$Sn_{pre} = Sn_{Ni3Sn} + Sn_{Ni3Sn2} + Sn_{Evap} \tag{15}$$

In each formulation, the ratio between Ni_3Sn and Ni_3Sn_2 almost matches the ratio between the intensities of the main peaks (R) in the XRD pattern, i.e., $R = Ni_{Ni3Sn}/Ni_{Ni3Sn2}$, then if there is no free Ni:

$$Ni_{Pre} = \frac{3M.wt_{Ni}}{M.wt_{Ni3Sn}} \times Ni3Sn + \frac{3M.wt_{Ni}}{M.wt_{Ni3Sn2}} \times Ni3Sn2 \tag{16}$$

$$Ni_{Pre} = \frac{3M.wt_{Ni}}{M.wt_{Ni3Sn}} \times R \times Ni3Sn2 + \frac{3M.wt_{Ni}}{M.wt_{Ni3Sn2}} \times Ni3Sn2 \tag{17}$$

$$Ni3Sn2 = \frac{Ni_{Pre}}{3M.wt_{Ni} \times \left(\frac{R}{M.wt_{Ni3Sn}} + \frac{1}{M.wt_{Ni3Sn2}} \right)} \tag{18}$$

With the same strategy, for full consumption of Sn in the formed alloys:

$$Sn_{Pre} = \frac{M.wt_{Sn}}{M.wt_{Ni3Sn}} \times Ni3Sn + \frac{2M.wt_{Sn}}{M.wt_{Ni3Sn2}} \times Ni3Sn2 \tag{19}$$

$$Sn_{Pre} = \frac{M.wt_{Sn}}{M.wt_{Ni3Sn}} \times R \times Ni3Sn2 + \frac{2M.wt_{Sn}}{M.wt_{Ni3Sn2}} \times Ni3Sn2 \tag{20}$$

$$Ni3Sn2 = \frac{Sn_{Pre}}{M.wt_{Sn} \times \left(\frac{R}{M.wt_{Ni3Sn}} + \frac{2}{M.wt_{Ni3Sn2}} \right)} \tag{21}$$

From these equations, the amount of Ni_3Sn and Ni_3Sn_2, and consequently Ni and Sn in each formulation can be estimated. Later on, the formed graphite can also be estimated from the TGA data. From the calculations, it was observed that the losses in tin occurred in the highest tin precursor sample (35 wt %), as Ni was totally consumed in the metallic alloy, so the remaining Sn evaporated and no free nickel was obtained. Numerically, the rough estimation of the elemental composition of the produced nanofibers is summarized in Table 1.

Table 1. Composition of the produced nanofibers produced at 850 °C.

SnCl$_2$ in the Initial Solution (%)	NiSn Alloys Content		Elemental Compostion				
	Ni$_3$Sn$_2$ (%)	Ni$_3$Sn (%)	Ni (in NiSn Alloys) (%)	Ni (Free) (%)	Sn (%)	Sn$_{evap}$ (%)	C (%)
5	42	58	5.457	28.082	4.510	0	61.952
10	100	0	6.544	24.380	8.779	0	60.297
25	59	41	23.399	0.459	20.31	0	55.823
35	54	46	19.703	0	16.890	10.215	53.192

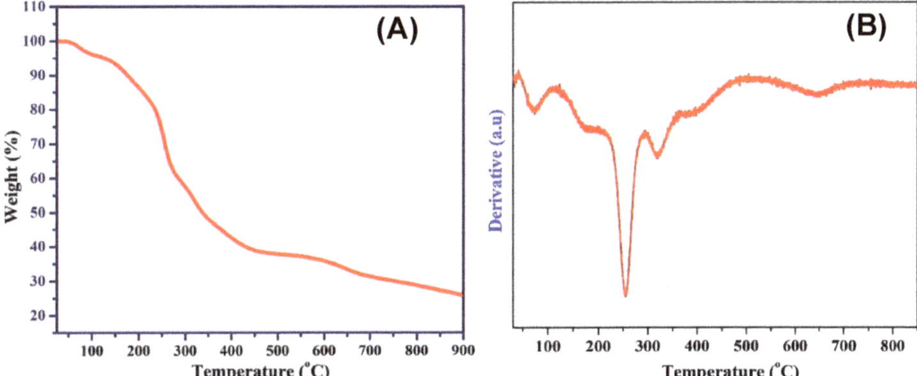

Figure 4. Thermal gravimetric analysis (TGA) (**A**), and first derivative for the TGA data (**B**) for an electrospun mat containing 10 wt % tin precursor.

2.2. Electrochemical Performance

ESA Investigation

Activation of the nickel-based electrodes can be carried in situ or in vivo. In other words, the electrode can be preliminarily activated by sweeping in a strong alkaline solution before use, or simultaneously during the electrooxidation process [23]. The activation denotes formation of ESA from the NiOOH active compound. During the activation, usually two regions are observed; the first one is seen in the negative potential region (at ~ −650 mV) due to formation of nickel hydroxide as follows [57]:

$$Ni + 2OH^- \rightarrow Ni(OH)_2 + 2e \tag{22}$$

The corresponding peak is usually very small in the first cycle and disappears in the subsequent ones [57–59]. The second transformation is done in the positive potential side associated with appearance of a strong peak related to the oxidation of Ni(OH)$_2$ to NiOOH according to the following reaction [59–61]:

$$Ni(OH)_2 + OH^- \rightarrow NiOOH + H_2O + e \tag{23}$$

Increasing the number of sweeping cycles leads to a progressive increase of the current density values of the cathodic peak due to the entry of OH$^-$ into the Ni(OH)$_2$ surface layer, which results in a progressive formation of a thicker NiOOH layer [57].

Figure 5A demonstrates the cyclic voltammograms for nanofibers with different tin contents and that were calcined at 700 °C. As shown, the $Ni(OH)_2/NiOOH$ transformation peaks are clearly observed for all formulations. The ESA can be calculated from the obtained graphs according to the following equation [62–64]

$$ESA = \frac{Q}{mq} \tag{24}$$

where Q is the charge required to reduce NiOOH to $Ni(OH)_2$, m is the loading amount of the catalysts, and q is the charge associated with the formation of a monolayer from $Ni(OH)_2$. As only one electron is needed to perform NiOOH to $Ni(OH)_2$ transformation, the value of q can be used as 257 µC/cm^2 [65,66]. Q can be determined from the area of the cathodic peak. Figure 5B displays the influence of Sn content on the ESA of the prepared nanofibers. From the results, it can be concluded that if Sn content was optimized, a distinct enhancement in the ESA can be performed. As shown, ESA of the nanofibers prepared from a solution containing 15 wt % tin precursor increased to 5 times higher than the Sn-free nanofibers; the estimated ESAs were 275 and 53 cm^2/mg for the two formulations, respectively. There are also other compositions of the nanofibers that reveal high ESA compared to the un-doped Ni/C ones. Typically, the ESAs created on the surface of nanofibers containing 10 and 25 wt % Sn are 110.2 and 87.7 cm^2/mg, which is 2 and 1.5 times higher than the Ni/C nanofibers, respectively. Other formulations (5 and 35 wt % Sn) reveal lower activity. This finding indicates that the alloy composition strongly affects the ESA formation on the surface of the electrocatalyst due to the change in the electronic composition.

Increasing the synthesis temperature to 850 °C leads to a considerable change in the obtained results, as shown in Figure 6. As can be observed, the optimum tin content is 10 wt %. At this composition, the ESA is almost 6 times more compared to Sn-free Ni/C nanofibers, at 175 and 30 cm^2/mg, respectively. Also, the nanofibers with 5 wt % Sn which displayed a low performance when they were synthesized at 700 °C, while at 850 °C these nanofibers became the second best among the investigated samples. Excluding the nanofibers containing the maximum Sn content, the remaining nanofibers showed a close ESA. Consequently, the synthesis temperature has an important impact on the ESA formation on the surface of the investigated nanofibers.

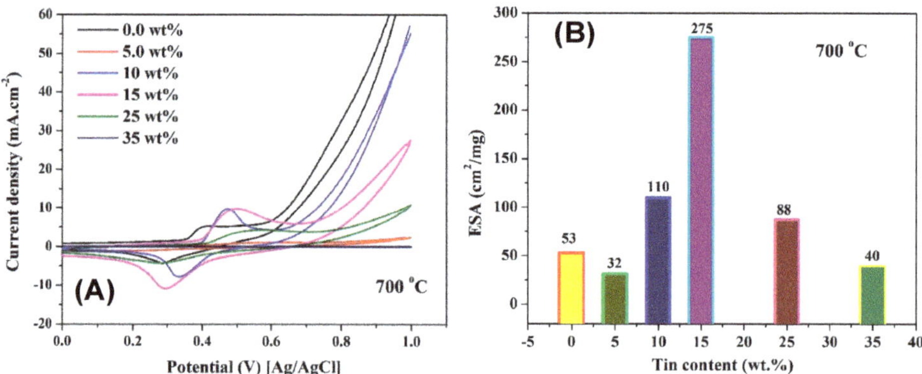

Figure 5. Activation of Sn-incorporated Ni/C nanofibers with different Sn contents and prepared at 700 °C (**A**); and influence of the Sn content on the electrochemical surface area (**B**).

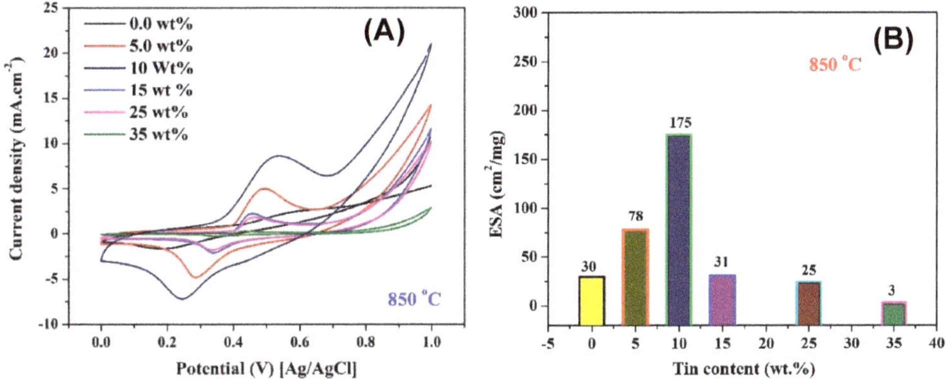

Figure 6. Activation of Sn-incorporated Ni/C nanofibers with different Sn contents and prepared at 850 °C (**A**), and influence of the Sn content on the electrochemical surface area (**B**).

To properly investigate the influence of the synthesis temperature, Figure 7 shows a comparison between three synthesis temperatures for nanofibers containing 10 wt % Sn. As can be concluded from the figure, further increase in the synthesis temperature (to 1000 °C) does have a negative influence on the area of the cathodic peak and consequently on the ESA. Numerically, the determined ESA formed on the surface of the nanofibers prepared at 700, 850, and 1000 °C is 110, 175, and 25 cm^2/mg, respectively. Paradoxically, the sample with 15 wt % co-catalyst reveals very high ESA when synthesized at 700 °C compared to 850 °C (Figures 5A and 6A). However, all the samples synthesized at 1000 °C revealed very low ESA values compared to the other two temperatures; data are not shown. Accordingly, it can be concluded that the optimum synthesis temperature is either 700 or 850 °C according to the nanofiber composition. Another important finding that can be observed from these results is the shift of the cathodic peak to the negative potential direction in the case of the nanofibers containing 10 wt % Sn and synthesized at 850 °C. This shift leads to a distinct decrease in the onset potentials of the electrooxidation process.

Electron transfer resistance is a very important parameter affecting the electrocatalyst performance. Accordingly, the catalyst morphology can have a considerable impact. Beside the Ohm resistance of the electrocatalyst, the electron faces another kind of resistance to transfer through the catalyst layer; the interfacial resistance. Therefore, the axial ratio is an important characteristic. In other words, the long axial ratio nano-morphology is the low interfacial resistance. In detail, in the case of the spherical nanoparticles, there are numerous contact points with very small contact areas, which creates a high interfacial resistance. On the other hand, in the case of the nanofibrous morphology, the long axial ratio leads to formulation of direct channels to the current collector, which strongly decreases the interfacial resistance. Figure 8A displays a schematic illustration for the electron that passes through the nanoparticles and the nanofibers. The average diameter of the fabricated nanoparticles was ~235 nm, while the composition was almost similar to the corresponding nanofibers. As shown, in the case of the nanoparticles, the electron has to pass through several contact points in a zigzag pass, which means high interfacial resistance. On the other hand, the long axial ratio helps to create direct passes for the electrons to reach to the current collector [44]. The aforementioned hypothesis was scientifically proved by synthesis nanoparticles from the same sol-gel used in the electrospinning process and calcined at the same temperature; 10 wt % Sn at 850 °C. As shown in Figure 8B, there is a big difference between the created ESA in the two formulations. Accordingly, it is expected that the nanofibrous morphology will strongly enhance the electrocatalytic activity.

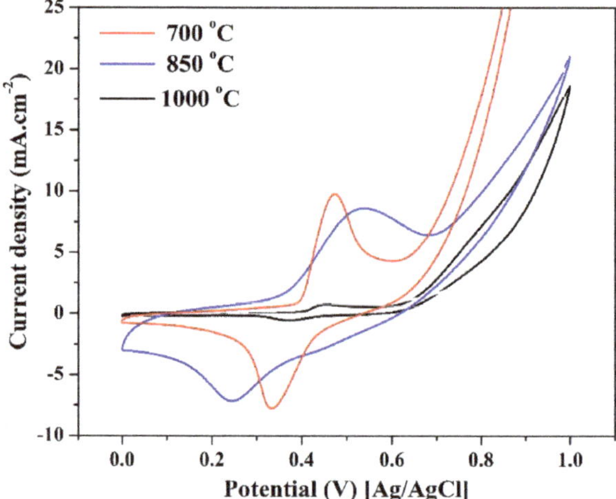

Figure 7. Influence of the calcination temperature on the ESA of the activated nanofibers containing 10 wt % Sn.

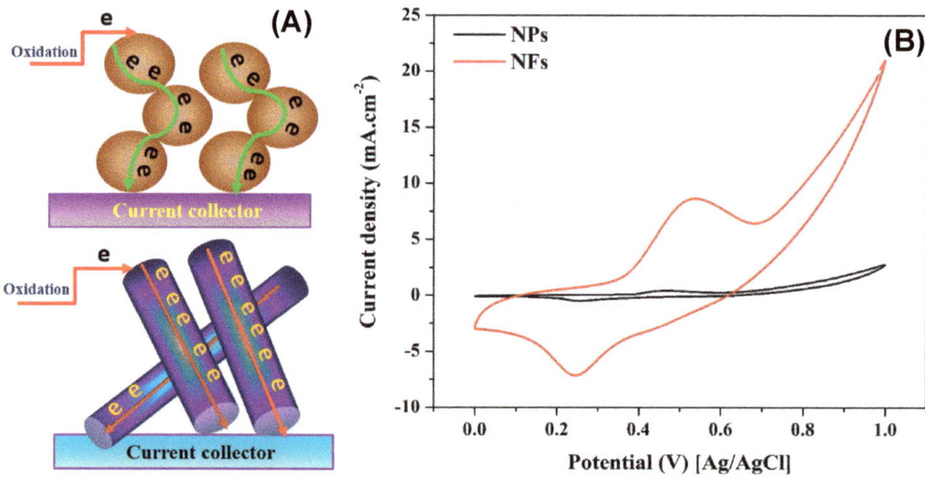

Figure 8. Schematic illustration of the electron passing through the nanoparticles and nanofibers (**A**). Influence of the nanostructural morphology on the ESA of the activated composites containing 10 wt % Sn and calcined at 850 °C (**B**).

To scientifically prove the influence of the ESA on the electrocatalytic activity of the proposed composites, the performance of the proposed materials toward the electrooxidation of methanol and urea was investigated. Figure 9 displays the electroactivity of the prepared nanofibers at different Sn contents prepared at 700 and 850 °C toward methanol oxidation (1.0 M methanol in 1.0 KOH, scan rate 50 mV/s). As shown in Figure 9A, the nanofibers owing the maximum ESA (15 wt % Sn and prepared at 700 °C) reveal the best activity. Similarly, in the case of the 850 °C calcination temperature, the best performance was associated with the maximum ESA, with the nanofibers containing 10 wt %. However, at the two temperatures, the performance does not follow the ESA order. For instance, for the two temperatures, the second best performance is associated with the Sn-free nanofibers. Moreover, for the remaining Sn-containing nanofibers, the electro-catalytic activity sequence does not follow the ESA order for the two temperatures. This result draws the attention toward an important conclusion, that NiOOH is not the main tool for the methanol oxidation. In other words, as was concluded in our previous study, methanol can be oxidized on the surface of the un-activated nickel electrode with simultaneous formation of NiOOH [23]. Therefore, the current study supports our previous one and introduces a new understanding for methanol oxidation. However, the two studies confirm that the NiOOH-based mechanism is essential but not unique. Beside this new finding, there is another one that can be extracted from Figure 9B. As shown, there is good enhancement in the methanol oxidation onset potential in the case of the nanofibers containing 10 wt % Sn and prepared at 850 °C. It is important to link with the negative shift observed during the activation of these nanofibers (Figure 6A). In more detail, the formation potential of the NiOOH affects the methanol oxidation onset potential. This finding is very important from an application point of view. Decreasing the onset potential is highly preferable in the case of production of hydrogen from methanol using the electrolysis process because it results in decreasing the required power.

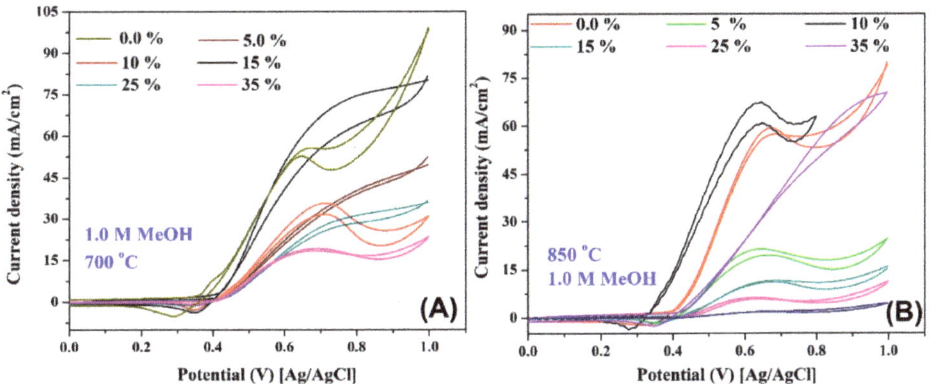

Figure 9. Electrocatalytic activity of Ni/C nanofibers containing different Sn contents and prepared at 700 °C (**A**) and 850 °C (**B**) toward methanol oxidation (1.0 M in 1.0 KOH) at room temperature and at a scan rate of 50 mV/s.

Figure 10 displays the influence of Sn content on the electrocatalytic activity of the proposed nanofibers toward urea oxidation; 1.0 M urea in 1.0 M KOH, scan rate 50 mV/s. As seen in the figure, for the two synthesis temperatures, Sn incorporation distinctly improves the electrocatalytic activity. In contrast to methanol oxidation, Sn-free Ni/C nanofibers reveal low electrocatalytic activity toward urea oxidation. Another important piece of information that can be gained from the figure is that the electrocatalytic activity almost follows the ESA sequence in the prepared nanofibers, which indicates that the urea oxidation mainly depends on the amount of the formed NiOOH molecules on the surface of the electrocatalyst. This result also supports the proposed urea oxidation mechanism, confirming the dependence on the NiOOH species [17,62]. Similar to methanol, the nanofibers containing 10 wt % Sn and prepared at 850 °C show very low onset potential for urea oxidation compared to the other formulations, as seen in Figure 10B. Numerically, the corresponding onset potential decreased to 195 mV (vs. Ag/AgCl), while all other formulations prepared at the same temperature (850 °C) have an onset potential of ~ 410 mV. This finding is interesting because it indicates the possibility of exploiting these nanofibers as an anode material in the direct urea fuel cells. Table 2 summarizes the onset potentials of urea and methanol oxidation at the surface of the proposed nanofibers.

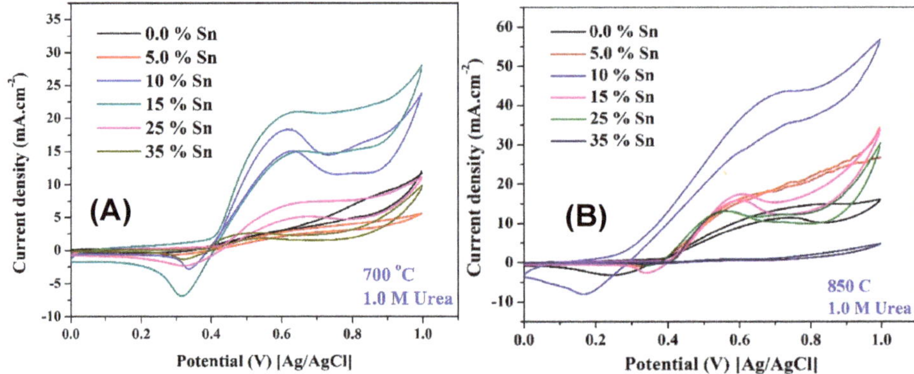

Figure 10. Electrocatalytic activity of Ni/C nanofibers containing different Sn contents and prepared at 700 °C (**A**) and 850 °C (**B**) toward urea oxidation (1.0 M in 1.0 KOH) at room temperature and a scan rate of 50 mV/s.

Table 2. The onset potentials (mV vs. Ag/AgCl) of urea and methanol oxidation at the surface of the proposed nanofibers.

	Synthesis Temp. °C	Sn Precusor Content (wt %)					
		0	5	10	15	25	35
Methanol	700	375	415	412	405	423	425
	850	395	403	360	415	445	415
Urea	700	408	412	405	401	415	425
	850	410	412	195	405	406	408

According to the XRD results, the interesting small onset-potential obtained with the 10 wt % sample synthesized at 850 °C, especially in urea oxidation, can be attributed to the formation of only Ni_3Sn_2 alloy with some free non-crystalline nickel (Table 1). Therefore, it can be claimed that this alloy structure possesses higher electrocatalytic activity toward urea oxidation compared to the Ni_3Sn one.

Figures 11 and 12 display the influence of the calcination temperature and the nano-morphology on the electrocatalytic activity of the proposed nanocomposites toward methanol and urea oxidation, respectively. As shown in Figure 11A, which displays the influence of the synthesis temperature on the electrocatalytic activity of the nanofibers containing 10 wt % Sn toward methanol, the optimum temperature is 850 °C, and increasing the temperature does have a negative impact on the activity. Moreover, the performance follows the ESA sequence. In the same fashion, as shown in Figure 11B, the electrocatalytic activity toward urea oxidation depends on the formed NiOOH active layer (ESA) on the surface of the investigated nanofibers.

Figure 12 displays a good translation for the results obtained in Figure 8 (influence of the nano-morphology on the ESA), and simultaneously supports the aforementioned hypothesis about the role of the axial ratio in enhancing the electron transfer process. As shown in the figure, for both methanol and urea, the nanofibrous morphology provides an advantage for the investigated electrocatalyst in enhancing the electrocatalytic activity toward oxidation of methanol (Figure 12A) and urea (Figure 12B).

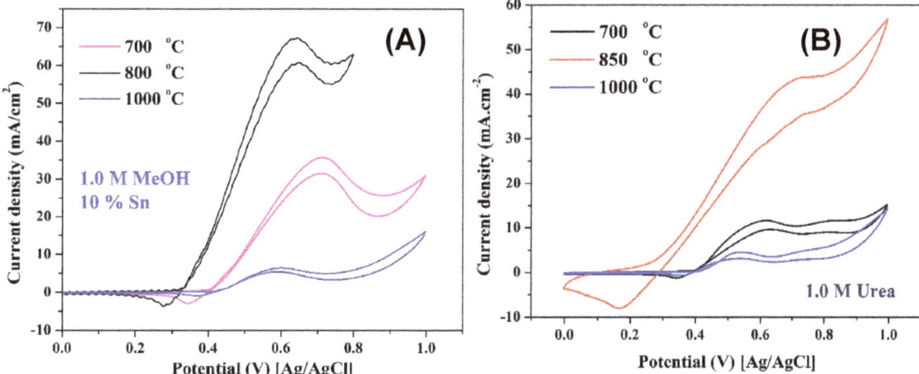

Figure 11. Influence of calcination temperature on the electrocatalytic activity of 10 wt % Sn nanofibers toward oxidation of methanol (1.0 M in 1.0 KOH) (**A**) and urea (1.0 M in 1.0 KOH) (**B**); the measurements were performed at room temperature and a scan rate of 50 mV/s.

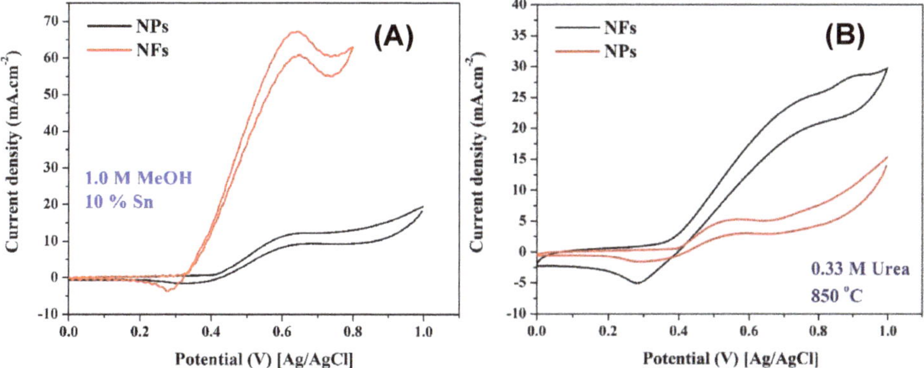

Figure 12. Influence of the nanostructural morphology on the electrocatalytic activity of 10 wt % Sn nanofibers prepared at 850 °C toward oxidation of methanol (1.0 M in 1.0 KOH) (**A**) and urea (1.0 M in 1.0 KOH); (**B**) the measurements were performed at room temperature and a scan rate of 50 mV/s.

3. Materials and Methods

3.1. Preparation of the Nanofibers and Nanoparticles

Sn-incorporated Ni/C nanofibers were prepared by dissolving (1 g) of nickel (II) acetate tetrahydrate (NiAc, 98%, Aldrich Co., Milwaukee, WI, USA) in 4 mL of de-ionized water that is mixed with 15 g of 10 wt % aqueous polymer solution of poly(vinyl alcohol) (PVA, MW = 65,000 g/mol, DC Chemical Co., Seoul, Korea). The mixture is stirred for several hours at 50 °C. Later on, specific amounts of tin chloride (SnCl$_2$; Aldrich Co., Milwaukee, WI, USA) were dissolved in the minimum amount of de-ionized water and mixed with the prepared NiAc/PVA solution. To study the influence of Sn content, solutions having 0, 5, 10, 15, 25, and 35 wt % of SnCl$_2$ compared to NiAc were prepared. For instance, a 10 wt % sample means the tin chloride was 10%, while the nickel acetate was 90 wt %. The electrospinning process was carried out at 20 kV at room conditions and at a 15 cm distance between the syringe and the rotating drum collector. After vacuum drying of the electrospun mats, the calcination process was carried out under vacuum atmosphere at different temperatures (700, 850, and 1000 °C) for 5 h. The nanoparticles were synthesized by drying sol-gels prepared by the aforementioned procedure under a vacuum for 24 h at 80 °C. Then, the solids were crushed and ground carefully into fine particles before being subjected to the calcination process.

3.2. Characterization

The nanofibrous morphology was confirmed by using scanning electron microscopy (SEM and FESEM, Hitachi S-7400, Tokyo, Japan). The chemical composition of the prepared nanostructures was studied by X-ray diffraction (XRD, Rigaku, Tokyo, Japan). Thermal properties have been studied by a thermal gravimetric analyzer (TGA, Pyris1, PerkinElmer Inc., Hopkinton, MA, USA). The electrochemical measurements were carried out using potentiostat (VersaStat 4, Princeton Applied Research. Co., Oak Ridge, TN, USA). A three electrode cell structure composed of glass carbon electrode (GCE) as the working electrode, Ag/AgCl as the reference electrode, and Pt wire as the counter electrode (CE) was utilized. WE was prepared by depositing 15 µL of catalyst ink onto the active surface of the GCE. The catalyst ink was prepared by well dispersion of the functional material in a mixture of Nafion solution and isopropanol followed by drying at 80 °C for 30 min.

4. Conclusions

Sn-incorporated Ni/C nanofibers can be produced by the electrospinning process. Due to the polycondensation characteristic of the main metal precursor (nickel acetate), well maintaining of the nanofibrous morphology during the calcination process can be achieved; accordingly, good morphology nanofibers are obtained. Addition of the proposed co-catalyst distinctly affects formation of the active NiOOH compound on the surface of the proposed electrocatalyst. However, the tin percentage should be optimized to maximize the electroactive surface area (ESA). Beside the composition, the synthesis temperature does have an important impact on the ESA. Typically, 700 and 850 °C calcination temperatures can be invoked based on the nanofiber composition. Due to the large axial ratio, which strongly assists the electron transfer process, the nanofibrous morphology distinctly maximizes ESA during the activation process. Although the electrocatalytic activity for the investigated nanofibers toward methanol oxidation does not exactly follow the ESA sequence, improving the ESA leads to a considerable decrease in the onset potential. On the other hand, the electrocatalytic activity toward urea oxidation is proportionate to the ESA, so the highest current density and lowest onset potential correspond to the nanofibers covered by the maximum ESA. Lastly, it is highly recommended to formulate the electrocatalysts in the nanofibrous morphology to gain the advantage of the large axial ratio, which distinctly improves the electocatalytic activity toward urea and methanol.

Author Contributions: N.A.M.B. planned the experimental work, wrote the manuscript, helped in the analysis, explained the results, and manipulated the data results. M.A.A. helped in the experimental work and writing of the manuscript. E.A.M.A. helped with the writing of the manuscript.

Funding: This research received no external funding.

Conflicts of Interest: The authors declare no conflict of interest.

References

1. Badwal, S.; Giddey, S.; Munnings, C.; Kulkarni, A. Review of progress in high temperature solid oxide fuel cells. *ChemInform* **2015**, *46*. [CrossRef]
2. Stambouli, A.B. Fuel cells: The expectations for an environmental-friendly and sustainable source of energy. *Renew. Sustain. Energy Rev.* **2011**, *15*, 4507–4520. [CrossRef]
3. Levin, D.B.; Pitt, L.; Love, M. Biohydrogen production: Prospects and limitations to practical application. *Int. J. Hydrog. Energy* **2004**, *29*, 173–185. [CrossRef]
4. Han, S.-K.; Shin, H.-S. Biohydrogen production by anaerobic fermentation of food waste. *Int. J. Hydrog. Energy* **2004**, *29*, 569–577. [CrossRef]
5. Zeng, K.; Zhang, D. Recent progress in alkaline water electrolysis for hydrogen production and applications. *Prog. Energy Combust. Sci.* **2010**, *36*, 307–326. [CrossRef]

6. Ni, M.; Leung, M.K.; Leung, D.Y.; Sumathy, K. A review and recent developments in photocatalytic water-splitting using TiO2 for hydrogen production. *Renew. Sustain. Energy Rev.* **2007**, *11*, 401–425. [CrossRef]

7. Grochala, W.; Edwards, P.P. Thermal decomposition of the non-interstitial hydrides for the storage and production of hydrogen. *Chem. Rev.* **2004**, *104*, 1283–1316. [CrossRef] [PubMed]

8. Barakat, N.A.; Abdelkareem, M.A.; Yousef, A.; Al-Deyab, S.S.; El-Newehy, M.; Kim, H.Y. Cadmium-doped cobalt/carbon nanoparticles as novel nonprecious electrocatalyst for methanol oxidation. *Int. J. Hydrog. Energy* **2013**, *38*, 3387–3394. [CrossRef]

9. Barakat, N.A.; El-Newehy, M.H.; Yasin, A.S.; Ghouri, Z.K.; Al-Deyab, S.S. Ni&Mn nanoparticles-decorated carbon nanofibers as effective electrocatalyst for urea oxidation. *Appl. Catal. A Gen.* **2016**, *510*, 180–188.

10. Barakat, N.A.; Motlak, M.; Elzatahry, A.A.; Khalil, K.A.; Abdelghani, E.A. NixCo1 − x alloy nanoparticle-doped carbon nanofibers as effective non-precious catalyst for ethanol oxidation. *Int. J. Hydrog. Energy* **2014**, *39*, 305–316. [CrossRef]

11. Demirbas, A. Biomethanol production from organic waste materials. *Energy Sources Part A* **2008**, *30*, 565–572. [CrossRef]

12. Amigun, B.; Gorgens, J.; Knoetze, H. Biomethanol production from gasification of non-woody plant in South Africa: Optimum scale and economic performance. *Energy Policy* **2010**, *38*, 312–322. [CrossRef]

13. Shamsul, N.; Kamarudin, S.K.; Rahman, N.A.; Kofli, N.T. An overview on the production of bio-methanol as potential renewable energy. *Renew. Sustain. Energy Rev.* **2014**, *33*, 578–588. [CrossRef]

14. Hosseini, M.G.; Momeni, M.M. UV-cleaning properties of Pt nanoparticle-decorated titania nanotubes in the electro-oxidation of methanol: An anti-poisoning and refreshable electrode. *Electrochim. Acta* **2012**, *70*, 1–9. [CrossRef]

15. Mehmood, A.; Scibioh, M.A.; Prabhuram, J.; An, M.-G.; Ha, H.Y. A review on durability issues and restoration techniques in long-term operations of direct methanol fuel cells. *J. Power Sources* **2015**, *297*, 224–241. [CrossRef]

16. Abdelkareem, M.A.; Allagui, A.; Tsujiguchi, T.; Nakagawa, N. Effect of the Ratio Carbon Nanofiber/Carbon Black in the Anodic Microporous Layer on the Performance of Passive Direct Methanol Fuel Cell. *J. Electrochem. Soc.* **2016**, *163*, F1011–F1016. [CrossRef]

17. Yan, W.; Wang, D.; Botte, G.G. Nickel and cobalt bimetallic hydroxide catalysts for urea electro-oxidation. *Electrochim. Acta* **2012**, *61*, 25–30. [CrossRef]

18. King, R.L.; Botte, G.G. Investigation of multi-metal catalysts for stable hydrogen production via urea electrolysis. *J. Power Sources* **2011**, *196*, 9579–9584. [CrossRef]

19. Wang, D.; Yan, W.; Vijapur, S.H.; Botte, G.G. Electrochemically reduced graphene oxide–nickel nanocomposites for urea electrolysis. *Electrochim. Acta* **2013**, *89*, 732–736. [CrossRef]

20. Barakat, N.A.; Amen, M.T.; Al-Mubaddel, F.S.; Karim, M.R.; Alrashed, M. NiSn nanoparticle-incorporated carbon nanofibers as efficient electrocatalysts for urea oxidation and working anodes in direct urea fuel cells. *J. Adv. Res.* **2019**, *16*, 43–53. [CrossRef]

21. Ferdowsi, G.S. Ni nanoparticle modified graphite electrode for methanol electrocatalytic oxidation in alkaline media. *J. Nanostruct. Chem.* **2015**, *5*, 17–23. [CrossRef]

22. Das, S.; Dutta, K.; Kundu, P.P. Nickel nanocatalysts supported on sulfonated polyaniline: Potential toward methanol oxidation and as anode materials for DMFCs. *J. Mater. Chem. A* **2015**, *3*, 11349–11357. [CrossRef]

23. Barakat, N.A.; Yassin, M.A.; Al-Mubaddel, F.S.; Amen, M.T. New electrooxidation characteristic for Ni-based electrodes for wide application in methanol fuel cells. *Appl. Catal. A Gen.* **2018**, *555*, 148–154. [CrossRef]

24. Pyun, S.-I.; Kim, K.-H.; Han, J.-N. Analysis of stresses generated during hydrogen extraction from and injection into Ni(OH)$_2$/NiOOH film electrode. *J. Power Sources* **2000**, *91*, 92–98. [CrossRef]

25. Klaus, S.; Cai, Y.; Louie, M.W.; Trotochaud, L.; Bell, A.T. Effects of Fe electrolyte impurities on Ni(OH)$_2$/NiOOH structure and oxygen evolution activity. *J. Phys. Chem. C* **2015**, *119*, 7243–7254. [CrossRef]

26. Guo, F.; Ye, K.; Cheng, K.; Wang, G.; Cao, D. Preparation of nickel nanowire arrays electrode for urea electro-oxidation in alkaline medium. *J. Power Sources* **2015**, *278*, 562–568. [CrossRef]

27. Ye, K.; Zhang, D.; Guo, F.; Cheng, K.; Wang, G.; Cao, D. Highly porous nickel@ carbon sponge as a novel type of three-dimensional anode with low cost for high catalytic performance of urea electro-oxidation in alkaline medium. *J. Power Sources* **2015**, *283*, 408–415. [CrossRef]

28. Barakat, N.A.M.; Moustafa, H.M.; Nassar, M.M.; Abdelkareem, M.A.; Mahmoud, M.S.; Almajid, A.A.; Khalil, K.A. Distinct influence for carbon nano-morphology on the activity and optimum metal loading of Ni/C composite used for ethanol oxidation. *Electrochim. Acta* **2015**, *182*, 143–155. [CrossRef]

29. Zhu, H.; Wang, J.; Liu, X.; Zhu, X. Three-dimensional porous graphene supported Ni nanoparticles with enhanced catalytic performance for Methanol electrooxidation. *Int. J. Hydrog. Energy* **2017**, *42*, 11206–11214. [CrossRef]

30. Danaee, I.; Jafarian, M.; Mirzapoor, A.; Gobal, F.; Mahjani, M.G. Electrooxidation of methanol on NiMn alloy modified graphite electrode. *Electrochim. Acta* **2010**, *55*, 2093–2100. [CrossRef]

31. Deng, Z.; Yi, Q.; Zhang, Y.; Nie, H. NiCo/C-N/CNT composite catalysts for electro-catalytic oxidation of methanol and ethanol. *J. Electroanal. Chem.* **2017**, *803*, 95–103. [CrossRef]

32. Tarrús, X.; Montiel, M.; Vallés, E.; Gómez, E. Electrocatalytic oxidation of methanol on CoNi electrodeposited materials. *Int. J. Hydrog. Energy* **2014**, *39*, 6705–6713. [CrossRef]

33. Gu, Y.; Luo, J.; Liu, Y.; Yang, H.; Ouyang, R.; Miao, Y. Synthesis of Bimetallic Ni–Cr Nano-Oxides as Catalysts for Methanol Oxidation in NaOH Solution. *J. Nanosci. Nanotechnol.* **2015**, *15*, 3743–3749. [CrossRef]

34. Wu, D.; Zhang, W.; Cheng, D. Facile Synthesis of Cu/NiCu Electrocatalysts Integrating Alloy, Core–Shell, and One-Dimensional Structures for Efficient Methanol Oxidation Reaction. *ACS Appl. Mater. Interfaces* **2017**, *9*, 19843–19851. [CrossRef]

35. He, H.; Xiao, P.; Zhou, M.; Zhang, Y.; Lou, Q.; Dong, X. Boosting catalytic activity with a p–n junction: Ni/TiO$_2$ nanotube arrays composite catalyst for methanol oxidation. *Int. J. Hydrog. Energy* **2012**, *37*, 4967–4973. [CrossRef]

36. Guo, Y.G.; Hu, J.S.; Zhang, H.M.; Liang, H.P.; Wan, L.J.; Bai, C.L. Tin/Platinum Bimetallic Nanotube Array and its Electrocatalytic Activity for Methanol Oxidation. *Adv. Mater.* **2005**, *17*, 746–750. [CrossRef]

37. Aricò, A.S.; Antonucci, V.; Giordano, N.; Shukla, A.K.; Ravikumar, M.K.; Roy, A.; Barman, S.R.; Sarma, D.D. Methanol oxidation on carbon-supported platinum-tin electrodes in sulfuric acid. *J. Power Sources* **1994**, *50*, 295–309. [CrossRef]

38. Veizaga, N.S.; Rodriguez, V.I.; Rocha, T.A.; Bruno, M.; Scelza, O.A.; de Miguel, S.R.; Gonzalez, E.R. Promoting Effect of Tin in Platinum Electrocatalysts for Direct Methanol Fuel Cells (DMFC). *J. Electrochem. Soc.* **2015**, *162*, F243–F249. [CrossRef]

39. Li, X.; Wei, J.; Chai, Y.; Zhang, S. Carbon nanotubes/tin oxide nanocomposite-supported Pt catalysts for methanol electro-oxidation. *J. Colloid Interface Sci.* **2015**, *450*, 74–81. [CrossRef]

40. Li, Y.; Liu, C.; Liu, Y.; Feng, B.; Li, L.; Pan, H.; Kellogg, W.; Higgins, D.; Wu, G. Sn-doped TiO$_2$ modified carbon to support Pt anode catalysts for direct methanol fuel cells. *J. Power Sources* **2015**, *286*, 354–361. [CrossRef]

41. Yi, Q.F.; Huang, W.; Yu, W.Q.; Li, L.; Liu, X.P. Fabrication of Novel Titanium-supported Ni-Sn Catalysts for Methanol Electro-oxidation. *Chin. J. Chem.* **2008**, *26*, 1367–1372. [CrossRef]

42. Barakat, N.A.; Al-Mubaddel, F.S.; Karim, M.R.; Alrashed, M.; Kim, H.Y. Influence of Sn content on the electrocatalytic activity of NiSn alloy nanoparticles-incorporated carbon nanofibers toward methanol oxidation. *Int. J. Hydrog. Energy* **2018**, *43*, 21333–21344. [CrossRef]

43. Li, J.; Luo, Z.; Zuo, Y.; Liu, J.; Zhang, T.; Tang, P.; Arbiol, J.; Llorca, J.; Cabot, A. NiSn bimetallic nanoparticles as stable electrocatalysts for methanol oxidation reaction. *Appl. Catal. B* **2018**, *234*, 10–18. [CrossRef]

44. Barakat, N.A.; Abdelkareem, M.A.; El-Newehy, M.; Kim, H.Y. Influence of the nanofibrous morphology on the catalytic activity of NiO nanostructures: An effective impact toward methanol electrooxidation. *Nanoscale Res. Lett.* **2013**, *8*, 402. [CrossRef] [PubMed]

45. Tao, S.L.; Desai, T.A. Aligned Arrays of Biodegradable Poly(ε-caprolactone) Nanowires and Nanofibers by Template Synthesis. *Nano Lett.* **2007**, *7*, 1463–1468. [CrossRef]

46. Chen, S. Self-Assembly of Perylene Imide Molecules into 1D Nanostructures: Methods, Morphologies, and Applications. *Chem. Rev.* **2015**, *115*, 11967–11998. [CrossRef]

47. Sehaqui, H.; Mushi, N.E.; Morimune, S.; Salajkova, M.; Nishino, T.; Berglund, L.A. Cellulose Nanofiber Orientation in Nanopaper and Nanocomposites by Cold Drawing. *ACS Appl. Mater. Interfaces* **2012**, *4*, 1043–1049. [CrossRef] [PubMed]

48. Barakat, N.A.M.; Abdelkareem, M.A.; Shin, G.; Kim, H.Y. Pd-doped Co nanofibers immobilized on a chemically stable metallic bipolar plate as novel strategy for direct formic acid fuel cells. *Int. J. Hydrog. Energy* **2013**, *38*, 7438–7447. [CrossRef]

49. Thamer, B.M.; El-Newehy, M.H.; Barakat, N.A.M.; Abdelkareem, M.A.; Al-Deyab, S.S.; Kim, H.Y. Influence of Nitrogen doping on the Catalytic Activity of Ni-incorporated Carbon Nanofibers for Alkaline Direct Methanol Fuel Cells. *Electrochim. Acta* **2014**, *142*, 228–239. [CrossRef]

50. Zhao, Y.; Yang, X.; Tian, J.; Wang, F.; Zhan, L. Methanol electro-oxidation on Ni@ Pd core-shell nanoparticles supported on multi-walled carbon nanotubes in alkaline media. *Int. J. Hydrog. Energy* **2010**, *35*, 3249–3257. [CrossRef]

51. Asgari, M.; Maragheh, M.G.; Davarkhah, R.; Lohrasbi, E.; Golikand, A.N. Electrocatalytic oxidation of methanol on the nickel–cobalt modified glassy carbon electrode in alkaline medium. *Electrochim. Acta* **2012**, *59*, 284–289. [CrossRef]

52. Zhong, J.-P.; Fan, Y.-J.; Wang, H.; Wang, R.-X.; Fan, L.-L.; Shen, X.-C.; Shi, Z.-J. Highly active Pt nanoparticles on nickel phthalocyanine functionalized graphene nanosheets for methanol electrooxidation. *Electrochim. Acta* **2013**, *113*, 653–660. [CrossRef]

53. Yu, M.; Chen, J.; Liu, J.; Li, S.; Ma, Y.; Zhang, J.; An, J. Mesoporous $NiCo_2O_4$ nanoneedles grown on 3D graphene-nickel foam for supercapacitor and methanol electro-oxidation. *Electrochim. Acta* **2015**, *151*, 99–108. [CrossRef]

54. Yousef, A.; Barakat, N.A.; Amna, T.; Unnithan, A.R.; Al-Deyab, S.S.; Kim, H.Y. Influence of CdO-doping on the photoluminescence properties of ZnO nanofibers: Effective visible light photocatalyst for waste water treatment. *J. Lumin.* **2012**, *132*, 1668–1677. [CrossRef]

55. Nash, P.; Nash, A. The Ni−Sn (Nickel-Tin) system. *Bull. Alloy Phase Diagr.* **1985**, *6*, 350–359. [CrossRef]

56. Yousef, A.; Barakat, N.A.M.; El-Newehy, M.; Kim, H.Y. Chemically stable electrospun NiCu nanorods@carbon nanofibers for highly efficient dehydrogenation of ammonia borane. *Int. J. Hydrog. Energy* **2012**, *37*, 17715–17723. [CrossRef]

57. Rahim, A.; Hameed, R.A.; Khalil, M. Nickel as a catalyst for the electro-oxidation of methanol in alkaline medium. *J. Power Sources* **2004**, *134*, 160–169. [CrossRef]

58. Hahn, F.; Beden, B.; Croissant, M.; Lamy, C. In situ UV visible reflectance spectroscopic investigation of the nickel electrode-alkaline solution interface. *Electrochim. Acta* **1986**, *31*, 335–342. [CrossRef]

59. Vuković, M. Voltammetry and anodic stability of a hydrous oxide film on a nickel electrode in alkaline solution. *J. Appl. Electrochem.* **1994**, *24*, 878–882. [CrossRef]

60. Fleischmann, M.; Korinek, K.; Pletcher, D. The oxidation of organic compounds at a nickel anode in alkaline solution. *J. Electroanal. Chem. Interfacial Electrochem.* **1971**, *31*, 39–49. [CrossRef]

61. Enea, O. Molecular structure effects in electrocatalysis–II. Oxidation of d-glucose and of linear polyols on Ni electrodes. *Electrochim. Acta* **1990**, *35*, 375–378. [CrossRef]

62. Yan, W.; Wang, D.; Botte, G.G. Electrochemical decomposition of urea with Ni-based catalysts. *Appl. Catal. B Environ.* **2012**, *127*, 221–226. [CrossRef]

63. Xia, B.Y.; Wang, J.N.; Wang, X.X. Synthesis and application of Pt nanocrystals with controlled crystallographic planes. *J. Phys. Chem. C* **2009**, *113*, 18115–18120. [CrossRef]

64. Wang, D.; Yan, W.; Vijapur, S.H.; Botte, G.G. Enhanced electrocatalytic oxidation of urea based on nickel hydroxide nanoribbons. *J. Power Sources* **2012**, *217*, 498–502. [CrossRef]

65. Machado, S.A.; Avaca, L. The hydrogen evolution reaction on nickel surfaces stabilized by H-absorption. *Electrochim. Acta* **1994**, *39*, 1385–1391. [CrossRef]

66. Brown, I.; Sotiropoulos, S. Preparation and characterization of microporous Ni coatings as hydrogen evolving cathodes. *J. Appl. Electrochem.* **2000**, *30*, 107–111. [CrossRef]

 *catalysts*

Article

One Simple Strategy towards Nitrogen and Oxygen Codoped Carbon Nanotube for Efficient Electrocatalytic Oxygen Reduction and Evolution

Jing Sun [1], Shujin Wang [1], Yinghua Wang [1], Haibo Li [1], Huawei Zhou [1], Baoli Chen [1], Xianxi Zhang [1], Hongyan Chen [2], Konggang Qu [1,*] and Jinsheng Zhao [1,*]

[1] School of Chemistry and Chemical Engineering, Shandong Provincial Key Laboratory/Collaborative Innovation Center of Chemical Energy Storage & Novel Cell Technology, Liaocheng University, Liaocheng 252059, China; sunjing201809@gmail.com (J.S.); wsj9898521@gmail.com (S.W.); wyh18463588761@163.com (Y.W.); lihaibo@lcu.edu.cn (H.L.); zhouhuawei@lcu.edu.cn (H.Z.); chenbaoli@lcu.edu.cn (B.C.); xxzhang3@126.com (X.Z.)
[2] Collaborative Innovation Center of Antibody Drugs, Institute of BioPharmaceutical Research, Liaocheng University, Liaocheng 252059, China; chenhongyan1123@163.com
* Correspondence: kgqu1985@gmail.com (K.Q.); j.s.zhao@163.com (J.Z.); Tel.: +86-635-853-9077 (K.Q.); +86-635-853-9607 (J.Z.)

Received: 31 December 2018; Accepted: 2 February 2019; Published: 6 February 2019

Abstract: The development of advanced electrocatalysts for oxygen reduction and evolution is of paramount significance to fuel cells, water splitting, and metal-air batteries. Heteroatom-doped carbon materials have exhibited great promise because of their excellent electrical conductivity, abundance, and superior durability. Rationally optimizing active sites of doped carbons can remarkably enhance their electrocatalytic performance. In this study, nitrogen and oxygen codoped carbon nanotubes were readily synthesized from the pyrolysis of polydopamine-carbon nanotube hybrids. Different electron microscopes, Raman spectra and X-ray photoelectron spectroscopy (XPS) were employed to survey the morphological and componental properties. The newly-obtained catalyst features high-quality nitrogen and oxygen species, favourable porous structures and excellent electric conductivity, and thus exhibits remarkably bifunctional oxygen electrode activity. This research further helps to advance the knowledge of polydopamine and its potential applications as efficient electrocatalysts to replace noble metals.

Keywords: N, O-codoping;   polydopamine;   oxygen reduction;   oxygen evolution; electrocatalysts; bifunctional

---

1. Introduction

Green and sustainable energy sources play a crucial role in addressing concerns about the global energy dilemma, pollution, and climate change [1–3]. Regenerative energy techniques involving fuel cells, metal–air batteries, and electrocatalytic hydrogen production have attracted significant interest as energy storage and conversion devices [4,5], which usually involve cathodic oxygen reduction reaction (ORR) and anodic oxygen evolution reaction (OER). Both ORR and OER contain complicated multi-electron transfer processes, which would result in sluggish reaction kinetics and consequently compromise the whole performance of the above energy devices. Electrocatalysts can drastically promote reaction rates and lower overpotentials. These features make electrocatalysts indispensable components of energy devices [6,7]. The noble metal-based materials including Pt and Ru are the most efficient electrocatalysts for ORR or OER, but their prohibitive cost and scarcity greatly impedes the proliferation of related energy technologies. In comparison, carbon materials possess a series of satisfactory features, including natural abundance, superior electrical conductivity, tailorable structures and components, and strong anticorrosion abilities, and thus have been intensively studied as electrocatalysts during the last decade [5,8,9].

Nonmetallic heteroatom doping has been regarded as a valid strategy to increase the electrocatalytic activity of carbon materials and even achieve a particular activity towards certain electrocatalytic reactions [10]. Therefore, a rationally componential optimization is of paramount importance to develop the desired electrocatalysts. N-doping can render adjacent C atoms positive-charged, due to the greater electronegativity of N, and can also maintain the intrinsic electronic structure on account of the presence of a lone pair of nitrogen electrons. Thus, N-doping has been the most widely used [10]. Moreover, different N species significantly affect the catalytic activity of N-doping carbons. For example, Ruoff et al. found that pyridinic N (p-N) mainly determines the ORR activity of N–carbon catalysts, while graphitic N (g-N) lowers the onset potential of ORR [11]. Dai et al. suggested that g-N is responsible for ORR activity and considered p-N as active sites for the OER [12]. Most recently, oxygen species were found to be able to remarkably improve the bifunctional ORR and OER performance of carbon materials. Zhu et al. revealed that epoxy and ketene oxygen moieties provide more active sites for ORR and OER [13], whereas Wang et al. considered –COOH groups as highly bifunctionally active sites [14,15]. Although the underlying mechanisms remain controversial, N- and O-doping indeed apparently boost the bifunctional performance of carbon materials in ORR and OER. Meanwhile, dual-doped carbons have exhibited remarkably better performance than their single-doped counterparts due to the synergistic coupling effect of compatible dopants [16,17]. Significantly, N-doping within a certain range can increase the electrical conductivity of carbons [18], while the introduction of an O species is detrimental to the inherent electron configuration of sp^2-carbons. Therefore, the effective integration of N and O dopants into the carbon matrix requires a delicate design to achieve optimal synergistic performance.

Recently, we have exploited polydopamine (PDA) as an excellent platform for doped carbons due to its unrivaled structural and componential virtues [17,19]. Robust adhesion can make PDA easily grow onto the surfaces of different solids and thus realize structural regulation [5,20]. Simultaneously, the flexible componential tunability by the post-modification of PDA facilitates the importation of multiple kinds of heteroatoms to the derived carbons [17]. The inherent N component of PDA has been utilized for the development of single N-doped and codoped carbons, with secondary dopants including B, P and S [17]. Notably, besides one nitrogen-containing group, each monomer of PDA also contains two intrinsic oxygenated groups, which can be readily available to construct the N, O-codoped carbons, but thus far, less attention has been paid to them. However, the large amount of oxygen species will certainly compromise the electrical conductivity of the resultant carbons. Carbon nanotubes (CNTs) have been widely recognized as a good electronic conductor, which can be introduced as a conductive substrate to assure a favorable electronic transfer of the integrated materials [21].

Herein, we propose a novel and simple approach to obtain N, O-codoping carbon nanotubes. PDA was first easily grafted onto the surface of multi-walled CNTs and then subjected to high-temperature carbonization. Compared to previous post-doping routes, this facile strategy can directly generate in situ and uniform N, O-codoping into the resultant carbon materials together with favorable componential and structural features. Specifically, the obtained N, O-codoped CNT (N, O-CNT) contains 2.3% nitrogen and up to 12.6% oxygen. Impressively, the inherent pyrrolic N of PDA can be completely converted into p-N and g-N, and the C=O and –COOH species dominate among different oxygen groups. All of those components are electrocatalytically active for ORR and OER. Additionally, the network structure formed by cross-stacked carbon nanotubes provide excellent electrical conductivity and smooth mass transportation. Accordingly, the newly developed N, O-CNT exhibits bifunctional activities for both ORR and OER with enhanced activity and excellent stability. Because of the robust adhesion of PDA, this facile and straightforward N, O-codoping method is significantly promising for the future energy conversion and storage applications.

2. Results

2.1. The Structural and Componential Characterization of the Catalysts

The morphologies and nanostructures of the newly-obtained N, O-CNT were firstly investigated by a scanning electron microscope (SEM) and transmission electron microscopy (TEM). As shown in Figure 1A and Figure S1, SEM images show cross-stacked CNT networks. However, as presented in the TEM images (Figure 1C), all these CNTs preserve the isolated single-tube structure. The magnified TEM images (Figure 1C and Figure S2) further manifest a layer of continuous carbon thin-film with ~2 nm-thick coating outside the lattice fringe of the CNTs, which can be ascribed to the carbonization of the uniform PDA wrapping. Notably, the kind of stacked networks formed by isolated CNTs can easily deposit onto the electrode surface with abundant intrinsic inner cavities and large pores between the CNTs, which are believed to be conducive to facile electrolyte and gas transport during the electrocatalytic processes. As shown in Figure 1D, the elemental mapping images of N, O-CNT suggest a homogeneous distribution of C, O, and N components, further indicating PDA evenly wrapped onto the surface of the CNTs. The defects and structural evolution of the CNT samples can be reflected through the intensity ratio of the D band to the G band (I_D/I_G) in the Raman spectra. As displayed in Figure 1E, the p-CNT has a small I_D/I_G ratio of 0.38, displaying an intact, pristine structure. After acid oxidation of p-CNT, ox-CNT shows a much higher I_D/I_G ratio of 1.05 arising from the oxygenated-group-enriched CNT surface. Finally, N, O-CNT can inherit a large amount of nitrogen and oxygen dopants from PDA, possibly resulting in its relatively high I_D/I_G ratio of 0.87, even after high-temperature pyrolysis. Nitrogen adsorption of N, O-CNT shows a large surface area of up to 165 m^2 g^{-1}. The existence of different mesopores and macropores can be verified by the pore size distribution curve (Figure 1F), which also agrees with the electron microscopic results. Specifically, the smaller mesopores with a size of around 3−10 nm correspond to the intrinsic hollow structures of nanotubes, while the larger pores with sizes of around 10−150 nm are credited to the voids of the cross-stacking CNT networks.

FTIR measurements were used to confirm the formation of CNT-PDA hybrids (Figure 2A). The p-CNT has no characteristic peak because of its pristine structure. The peaks of ox-CNT at around 1050 and 1732 cm^{-1} indicated the presence of C–O and C=O moieties. The indole or indoline structure of PDA in CNT-PDA hybrids was certified by the coexistence of characteristic peaks at 1501 and 1609 cm^{-1}, indicating the successful wrapping of PDA onto the surface of CNT [22]. The chemical status of different components can be clearly determined by the XPS technique (Figure 2B–D and Figure S3). The presence of C, N (2.3 at.%) and O (12.6 at.%) was first confirmed by an XPS survey scan of N, O-CNT, in which the high concentration of N and O dopants also accounts for the high I_D/I_G ratio of N, O-CNT. Additionally, the deconvoluted spectra of N 1s consists of two peaks at 398.4 and 400.8 eV (Figure 2C), belonging to p-N (0.9 at.%), and g-N (1.4 at.%), while the catalytically inert pyrrolic N was not detected. The well-resolved O 1s peaks mainly comprise three peaks. More specifically, the peak at 532.5 eV is consistent with the hydroxyl (C−OH) and carbonyl (C=O) functional groups, which have a content of 5.6%. The peak at 531.3 eV is attributed to the carboxyl group (COO−) in carboxylate and the oxygen double bond to carbon, accounting for 1.7 at.%, and the peak at 533.9 eV corresponds to the oxygen single bond in esters and carboxylic acids (O=C−O), reaching 5.3% (Figure 2D). Significantly, compared with the oxygen content in p-CNT (2.3 at.%) and ox-CNT (9.5 at.%), shown in Figure S3, Figure S4 and Figure S5 [23], the relatively large number of oxygen moieties in N, O-CNT not only plays a critical role as active sites of electrocatalytic reactions [24–26], but also improves the wettable capability of the surface of resultant carbons and facilitates the electrocatalytic processes [27].

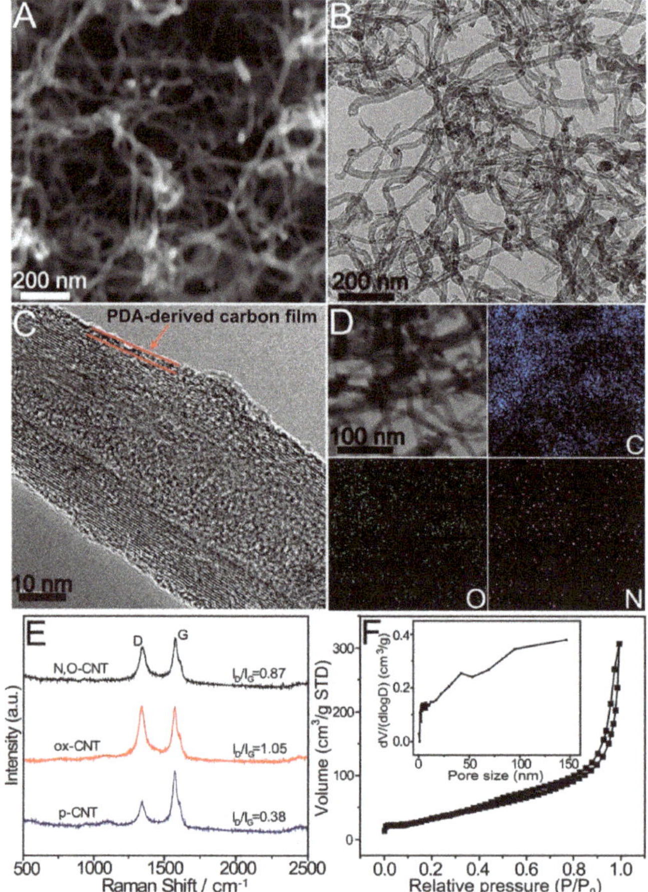

Figure 1. (**A**) SEM, (**B**) TEM and (**C**) the magnified TEM images of the N, O-carbon nanotube (CNT); PDA, polydopamine. (**D**) TEM elemental mapping of C, O, and N in N, O-CNT. (**E**) Raman spectra of p-CNT, ox-CNT and N, O-CNT. (**F**) The corresponding nitrogen adsorption–desorption isotherm of N, O-CNT, and the inset shows the pore size distribution curve.

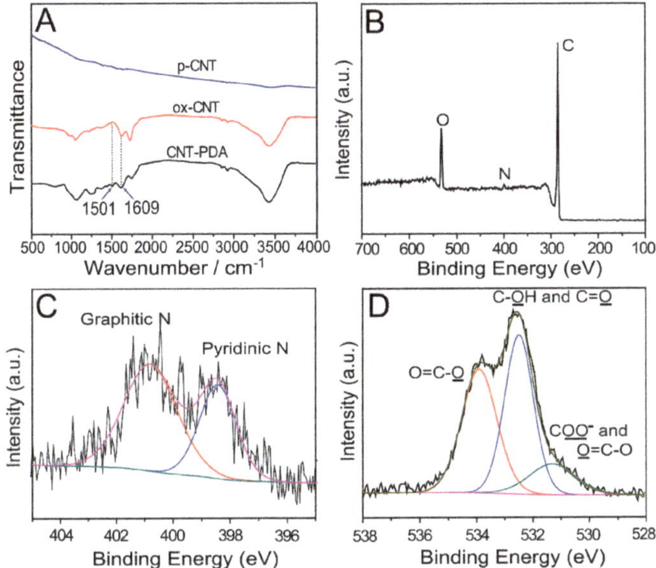

Figure 2. (**A**) FTIR spectra of p-CNT, ox-CNT and CNT-PDA. (**B–D**) XPS survey scan and the deconvoluted high-resolution spectra of N 1s and O 1s in N, O-CNT.

2.2. ORR Performance

The electrocatalytic activity for ORR was systematically evaluated on the newly-developed carbons. Firstly, the cyclic voltammograms (CVs) were examined in an O_2-saturated 0.1 M KOH electrolyte. Figure 3A displays N, O-CNT having a typical cathodic peak located at -0.28 V, and the peak potential is obviously much more positive than that of ox-CNT (-0.34 V) and p-CNT (-0.38 V). Meanwhile, the according peak current density of N, O-CNT is 3.3 mA cm^{-2}, much larger than that of ox-CNT (-0.94 mA cm^{-2}) and p-CNT (-1.54 mA cm^{-2}), both of which indicate the best ORR activity on N, O-CNT. Linear sweep voltammograms (LSVs) were then collected with a rotating disk electrode (RDE) under 1600rpm. The onset potential (η) was defined as the potential value to achieve a current density of -0.5 mA cm^{-2}. The η value of N, O-CNT is around -0.16 V, as shown in Figure 3B—much better than that of ox-CNT (-0.23 V) and p-CNT (-0.25 V). Of particular note, at -0.6 V, a current density up to -6.6 mA cm^{-2} can be observed on N, O-CNT, which is significantly larger than the values of ox-CNT (-2.4 mA cm^{-2}), p-CNT (-2.6 mA cm^{-2}), and even Pt/C (-4.6 mA cm^{-2}), suggesting superior electrocatalytic performance of N, O-CNT for ORR. Furthermore, N, O-CNT has no current plateau akin to that of Pt/C, caused by the limited mass transport, partly because the high surface area and hierarchically porous structures endow the N, O-CNT with abundantly available active sites and expedite diffusion ability in ORR electrocatalytic processes. The ORR catalytic kinetics of different catalysts were then assessed through Tafel slopes extracted from the LSV curves. Figure 3C illustrates a Tafel slope of 80 mV dec^{-1} on N, O-CNT, which is much smaller than that of p-CNT (98 mV dec^{-1}) and ox-CNT (124 mV dec^{-1}); nearly approaching the value of Pt/C (74 mV dec^{-1}), the small Tafel slope of N, O-CNT manifests the enhanced ORR kinetics of N, O-CNT. To gain more insight into the electrocatalytic ability of N, O-CNT, LSVs at different rotation speeds were recorded (Figure 3D and Figure S6). The respective Koutecky-Levich (KL) plots under various potentials were linearly fitted and the kinetic limiting current density (J_k) was obtained (Figure 3E). As presented in Figure 3F, N, O-CNT has relatively high and stable J_k values under the applied potential (-0.4 to -0.8 V) compared to those of ox-CNT and p-CNT, indicating a more efficient and smooth catalytic process for ORR on N, O-CNT due to its structural advantages, as noted previously.

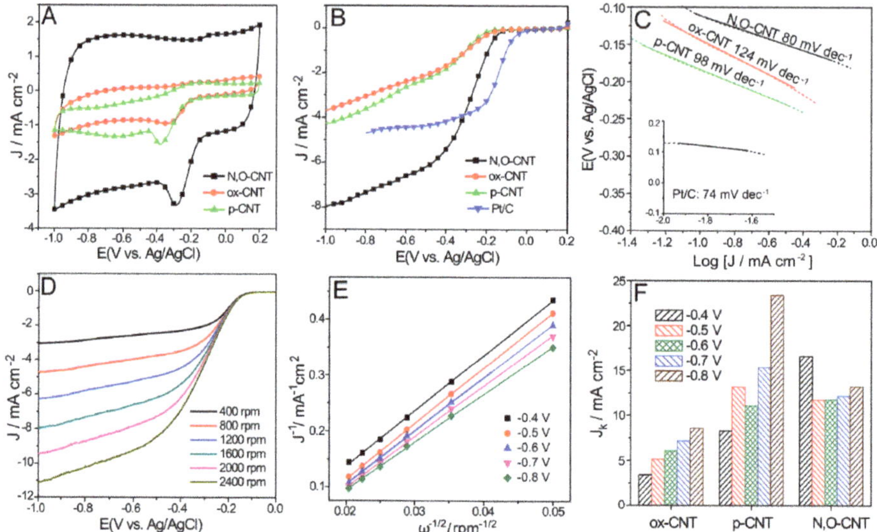

Figure 3. (**A**) CV curves of p-CNT, ox-CNT, and N, O-CNT in O_2-saturated 0.1 M KOH solution. (**B**) LSVs at a sweep rate of 5 mVs^{-1}, and (**C**) Tafel slope obtained from the LSVs of p-CNT, ox-CNT, N, O-CNT, and Pt/C. (**D**) Linear sweep voltammograms (LSVs) of N, O-CNT at different rotating speeds from 400 to 2400 rpm. (**E**) K-L plots of N, O-CNT obtained at different potentials. (**F**) Kinetic limiting current density (J_k) of different catalysts at a potential range from −0.4 to −0.8 V.

Rotating ring disk electrode (RRDE) experiments were further conducted to quantitatively analyze the intermediate peroxide and investigate the ORR pathways on different catalysts [28]. As presented in Figure 4A,B, N, O-CNT yields 17−34% HO_2^- with a potential from −0.4 to −1.0 V, and its number of electron transfer (n) ranges from 3.3 to 3.7. In comparison, p-CNT produces ∼41–5% HO_2^- under identical conditions, with n ranging from 2.5 to 3.2. The n of ox-CNT was estimated from 2.3 to 3.1, with the production of ∼44–84% HO_2^-. These results suggest a more efficient electrocatalytic ORR process on N, O-CNT with a 4e$^-$ dominated pathway. It is noteworthy that peroxide species produced in the low potential region can be continuously reduced at a high potential and contribute to the high reduction current density of N, O-CNT displayed in Figure 3B.

The long-term durability of N, O-CNT was assessed against commercial Pt/C. The test was conducted with chronoamperometry in 0.1 M KOH saturated with O_2. As displayed in Figure 4C, Pt/C exhibits up to a 40% loss from its initial current, while N, O-CNT retains 96.7% of the original current over 20 h with a neglectable attenuation, clearly manifesting the exceedingly good stability of carbon active sites in alkaline ORR. Furthermore, the effect of methanol crossover was investigated on both N, O-CNT and Pt/C (Figure 4D). After introducing methanol into the electrolyte, the original ORR current of N, O-CNT could persist almost unaffected, confirming its robust resistance to methanol crossover. Contrarily, when 3 M methanol was added, a quick response was detected on Pt/C with the initially cathodic current directly changing to an anodic current. Consequently, the obtained N, O-CNT showed prominent durability and high selectivity to ORR, and is highly suitable as a potential candidate to replace Pt/C.

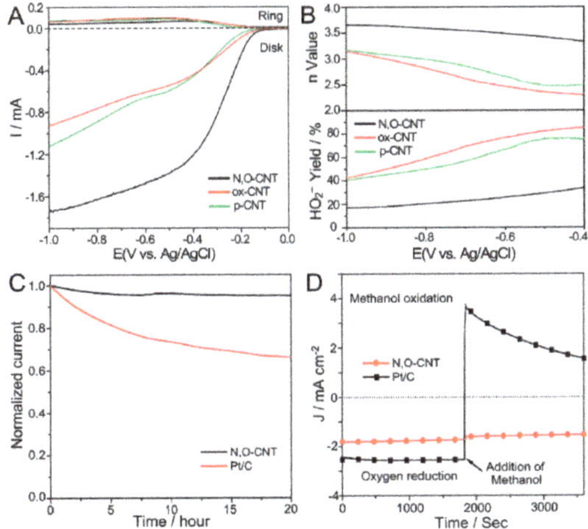

Figure 4. (**A**) Rotating ring disk electrode (RRDE) voltammetric response in the O_2-saturated 0.1 M KOH at a scan rate of 5 mV s^{-1} and (**B**) HO_2^- yields and the corresponding electron transfer number of p-CNT, ox-CNT and N, O-CNT. (**C**) oxygen reduction reaction (ORR) current–time chronoamperometric response of N, O-CNT and Pt/C in O_2-saturated 0.1 M KOH solution. (**D**) current–time chronoamperometric response of N, O-CNT and Pt/C before and after addition of 3 M methanol.

2.3. OER Performance

The OER catalytic performance of the obtained samples was further characterized in detail. LSVs were first performed in 0.1 M KOH, and the applied potential when generating a current density of 10 mA cm^{-2} ($E_{j=10}$) was a metric used for the comparison of different catalysts (Figure 5A). The $E_{j=10}$ for N, O-CNT is 0.65 V, which is much lower than those obtained for other samples, such as p-CNT (0.73 V) and ox-CNT (0.82 V), close to the value of IrO_2-CNT (0.61 V). Compared to previously reported catalysts, the E_j of N, O-CNT (0.65 V) is lower than that of the various carbons, including N, O-doped carbon hydrogels [21] and N-doped carbon nanocables [29], and comparable to those of metal-containing eletrocatalysts, such as $Mn_3O_4/CoSe_2$ hybrids [30], Co_3O_4/N-graphene [31], and Mn_xO_y/N-doped carbon [32]. Tafel plots were also used to evaluate the catalytic kinetics for OER (Figure 5B). N, O-CNT has a Tafel slope of 74 mV dec^{-1}, which is lowest among all the samples including p-CNT (93 mV dec^{-1}), ox-CNT (147 mV dec^{-1}), and even IrO_2-CNT (82 mV dec^{-1}). Compared with previously reported OER catalysts, the Tafel slope of N, O-CNT is much lower than those of N, O-doped carbon hydrogels (141 mV dec^{-1}) [21], C_3N_4/carbon nanotube composites (83 mV dec^{-1}) [33], and similar to some metal oxide OER catalysts, including Co_3O_4/carbon nanowires (70 mV dec^{-1}) [34], CoO/N-graphene (71 mV dec^{-1}) [35], and Co_3O_4/N-graphene (67 mV dec^{-1}) [31], implying its enormously beneficial catalytic kinetics for OER. The catalytic kinetics of different samples can be further evidenced by the electrochemical impedance spectrum (EIS). As illustrated in Figure 5C, N, O-CNT has a much smaller impedance compared to that of p-CNT and ox-CNT, confirming its greatly unimpeded reaction kinetics. An electrochemical durability test of N, O-CNT for OER was then carried out. As illustrated in the inset of Figure 4D, a long-time potential cycling conducted on N, O-CNT signified insignificant reduction of the catalytic performance after 1000 cycles (Figure 5D). The LSV curves show that 91.2% of the initial current density remained after 1000 potential cycles, confirming the remarkable electrochemical stability of the N, O-CNT for OER.

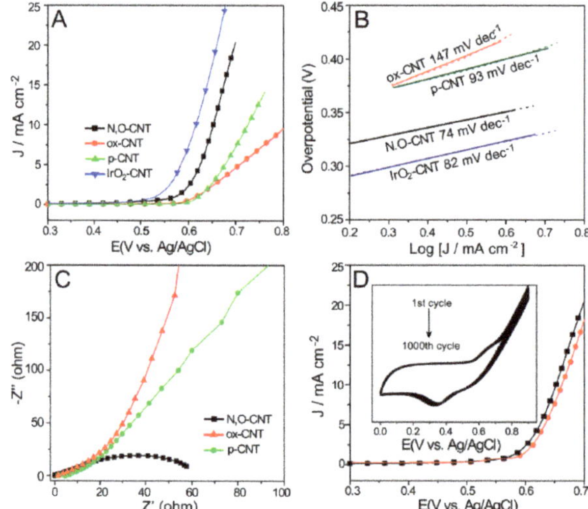

Figure 5. (**A**) Oxygen evolution reaction (OER) LSVs at a sweep rate of 5 mVs^{-1} and (**B**) OER Tafel plots of p-CNT, ox-CNT, N, O-CNT, and IrO$_2$-CNT. (**C**) The electrochemical impedance spectra (recorded at 0.65 V) of p-CNT, ox-CNT, and N, O-CNT. (**D**) Electrochemical durability test of N, O-CNT for OER, the LSV plots before and after 1000 cycles, and inset are the CV plots at 50 mV s^{-1} for 1000 cycles. (**A**) The overall LSV curve of N, S-CN in the potential range of −0.8 to 0.8V, ΔE (E$_{j=10}$ − E$_{j=−3}$) is a metric for bifunctional ORR and OER activity (Inset: The value of ΔE for various catalysts reported previously).

To better investigate the overall oxygen electrode activity, the different metrics for OER and ORR on various catalysts were all compared and displayed in Figure 6A and Table S1, including onset potential, Tafel slope, the potential at −3 mA cm^{-2} for ORR (E$_{j=−3}$), and E$_{j=10}$ for OER [36]. The difference in potential between E$_{j=10}$ and E$_{j=−3}$ was designated as ΔE, i.e., ΔE = E$_{j=10}$ − E$_{j=−3}$. The smaller ΔE means better overall oxygen electrode activity. Notably, N, O-CNT displays a ΔE of 0.92 V, much smaller than that of p-CNT (1.40 V) and ox-CNT (1.61 V), which is also superior compared to the previously reported non-metallic materials (e.g., N-graphene/CNT [37], ΔE = 0.96 V; C$_3$N$_4$-CNT, ΔE = 1.30 V) [36], noble-metals (e.g., Pt/C, ΔE = 1.16 V; Ru/C, ΔE = 1.01 V; Ir/C, ΔE = 0.92 V) [38], and transition-metals (e.g., NiCo$_2$S$_4$@N/S-rGO [39], ΔE = 0.98 V; NiCo$_2$O$_4$/G [40], ΔE=1.13 V) [32,38], and close to that of phosphorus-doped carbon nitride/carbon-fiber paper (PCN-CFP, 0.91 V) [36], N, S-CN (0.88 V) [6], and Co/N-C-800 (0.86 V) [41]. Figure 6B summarizes a detailed comparison of different bifunctional oxygen electrocatalysts, demonstrating the excellent catalytic performance of N, O-CNT towards a bifunctional ORR and OER.

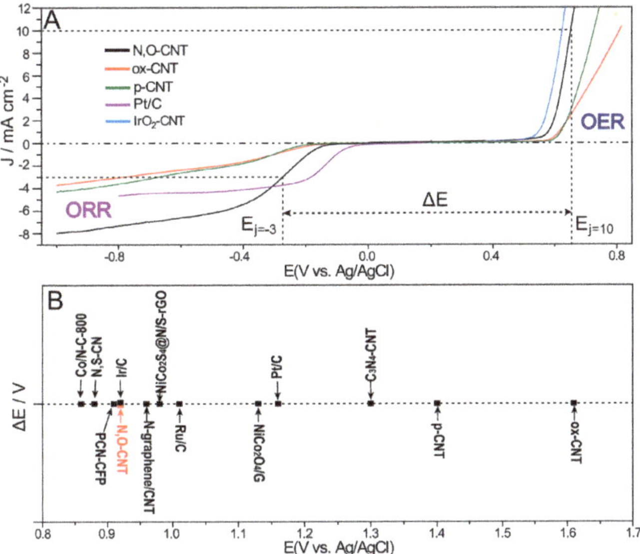

Figure 6. (**A**) The overall LSV curve of N, O-CNT in the potential range of −1.0 to 0.8 V, (**B**) the value of ΔE of N, O-CNT and the comparison with various catalysts reported previously.

The efficiently bifunctional performance of the N, O-CNT catalyst for ORR and OER could arise from the following three aspects: First, N, O-CNT contains 2.3% N and 12.6% O, and thus produces a large population of active sites. The N element has a larger electronegativity and can afford a positive charge density to the adjacent C atoms, which are generally considered electrocatalytically active centers [33,42]. Furthermore, N, O-CNT only consists of favored p-N and g-N, while no pyrrolic N was found, which reportedly has little catalytic effect [11]. Meanwhile, different oxygen groups, including the C=O and COOH moieties, have been found to facilitate OER and ORR due to the electron withdrawing effect and enhanced adsorption of reaction intermediates [14,15,43]. A host of oxygen groups up to 12.6% in N, O-CNT would, therefore, greatly promote the bifunctional activity of obtained carbons. Secondly, a large surface area of catalysts can enhance the exposure of active sites and assure their sufficient utilization. The porous structure of N, O-CNT, shown by nitrogen adsorption, can also be assessed by electrochemical double-layer capacitance (C_{dl}). Figure 7 displays the C_{dl} of N, O-CNT as 4.4 mF cm^{-2}, while the values of p-CNT and ox-CNT are 1.4 and 2.5 mF cm^{-2}. As C_{dl} reflects the electrocatalytic active surface area, the bigger C_{dl} of N, O-CNT illustrates the larger active surface of N, O-CNT, which can promote its apparent ORR and OER performance. Moreover, the different porous structures of N, O-CNT can guarantee an unblocked channel for benign mass transfer [22]. Thirdly, the employed CNT substrate can provide excellent electronic conductivity, which has been confirmed by the EIS study. Additionally, PDA-derived defective carbons can be firmly incorporated with CNTs because of the robustly adhesive PDA, which ensures an unimpeded charge transfer and the long-term stability of the integrated carbons. All these manifold virtues come together to make the developed N, O-CNT an advanced bifunctional ORR and OER electrocatalyst.

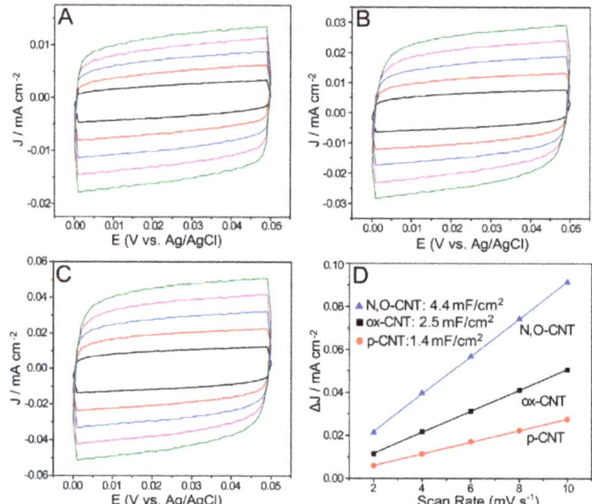

Figure 7. CV curves at different scan rates (2, 4, 6, 8, and 10 mV s^{-1}) of (**A**) p-CNT, (**B**) ox-CNT, and (**C**) N, O-CNT. (**D**) The corresponding difference in the current density at 0.025 V plotted against scan rate; the calculated C$_{dl}$ values are shown as the inset.

3. Materials and Methods

3.1. Preparation of N, O-CNT

The ox-CNT was synthesized first. The purchased primitive CNTs were ultrasonicated in a mixed solution of sulfuric acid (98%) and nitric acid (70%) with 3:1 v/v for 10 h and washed repeatedly with copious water. The obtained sample was incubated with 5 M HCl at 50 °C for 24 h to eliminate metal impurities. After lyophilization, ox-CNT was sonicated in water to obtain 1 mg mL^{-1} dispersion.

To prepare N, O-CNT, 100 mg of dopamine (DA) was added into 100 mL of the above ox-CNT dispersion followed by 100 mL of phosphate buffered saline (PBS, 0.4 M, pH = 8.5) added. Under a magnetic stirring, the reaction in the mixed solutions was kept for 24 h. The CNT-PDA samples were obtained by centrifugation and washing with water. The N, O-CNT was synthesized by the pyrolysis of CNT-PDA in a tube furnace under a N$_2$ atmosphere. The pyrolysis temperature was first set at 400 °C for 2 h with a heating rate of 1 °C min^{-1}, then at 800 °C for 3 h with a heating rate of 5 °C min^{-1}.

3.2. Electrochemical Characterization

The electrochemical measurements were conducted on an electrochemical workstation (CHI 760C, CH Instruments, Austin, TX, USA). The inks of different catalysts were prepared as follows: The catalysts of 2 mg were dispersed under ultrasonication into 1 ml water to make a well-distributed suspension. Then, 20 µL of catalyst ink was dropped on the electrode surface. 5 µL of 0.5 wt.% Nafion aqueous solution was pipetted on the electrode and air dried. The three-electrode cell system was employed in a standard five-neck electrolyzer and consisted of an RDE glassy carbon working electrode, a Ag/AgCl reference electrode in saturated AgCl-KCl solution, and a platinum wire as counter electrode. Cyclic voltammogram (CV) and linear sweep voltammogram (LSV) tests were performed with a scan rate of 50 and 5 mV s^{-1}, respectively. The RRDE measurement was conducted to evaluate the catalytic efficiency of samples for ORR, and its detailed experiments are presented in the Supplementary Materials.

The EIS tests of the OER were conducted under an AC voltage with 5 mV amplitude in a frequency range from 100,000 to 1 Hz and recorded at 0.65 V vs. Ag/AgCl. The C_{dl} of the as-synthesized materials was obtained from double-layer charging curves using CVs in a potential range of 0–0.05 V. The capacitive currents, i.e., $\Delta J_{|Ja-Jc|}$ @ 0.025 V, were plotted as a function of the CV scan rate. The linear relationship was observed with a slope two-times larger than the C_{dl} value.

4. Conclusions

In conclusion, by virtue of PDA, a new and simple strategy, with attractive componental and structural features, was presented to prepare N, O-CNT. The resultant codoped carbon is characterized by highly efficient N and O components, favorable pore architecture, and high surface areas, and hence exhibits a remarkably bifunctional performance for ORR and OER with outstanding activity and excellent stability. Due to the versatile features of PDA, this work could offer a novel insight into rationally developing PDA-derived doped carbons, which are greatly promising as substitutes for noble metals in relevant energy conversion fields.

Supplementary Materials: The following are available online at http://www.mdpi.com/2073-4344/9/2/159/s1, Figure S1: SEM image of N, O-CNT, Figure S2: The magnified TEM images of the N, O-CNT, Figure S3: XPS high-resolution spectra of C1s of N, O-CNT, Figures S4 and S5: XPS survey scans and the deconvoluted high-resolution spectra of p-CNT and ox-CNT. Figure S6: LSVs at different rotating speeds from 0 to 2400 rpm and the K-L plots obtained at different potentials, Table S1: Comparison of the different OER and ORR metrics of the obtained catalysts.

Author Contributions: Conceptualization, K.Q.; Methodology, J.S. and S.W.; Formal analysis, Y.W. and H.L.; Data curation, H.Z. and B.C.; Writing—original draft preparation, J.S. and H.C.; Writing—review and editing, K.Q. and J.Z.; Supervision, X.Z. and K.Q.; Project administration, K.Q.

Acknowledgments: This work was financially supported by the National Natural Science Foundation of China (21601078, 21503104), Natural Science Foundation of Shandong Province (ZR2016BQ21, ZR2014BQ010, ZR2016BQ20), Colleges and Universities in Shandong Province Science and Technology Projects (J16LC03, J16LC05, J17KA097), and the Doctoral Program of Liaocheng University (318051608).

Conflicts of Interest: The authors declare no conflict of interest.

References

1. Dresselhaus, M.S.; Thomas, I.L. Alternative energy technologies. *Nature* **2001**, *414*, 332–337. [CrossRef] [PubMed]
2. Norskov, J.K.; Christensen, C.H. Toward Efficient Hydrogen Production at Surfaces. *Science* **2006**, *312*, 1322–1323. [CrossRef] [PubMed]
3. Chu, S.; Cui, Y.; Liu, N. The path towards sustainable energy. *Nat. Mater.* **2017**, *16*, 16–22. [CrossRef] [PubMed]
4. Steele, B.C.H.; Heinzel, A. Materials for fuel-cell technologies. *Nature* **2001**, *414*, 345. [CrossRef] [PubMed]
5. Zhang, Q.; Luo, F.; Ling, Y.; Guo, L.; Qu, K.; Hu, H.; Yang, Z.; Cai, W.; Cheng, H. Constructing Successive Active Sites for Metal-free Electrocatalyst with Boosted Electrocatalytic Activities Toward Hydrogen Evolution and Oxygen Reduction Reactions. *ChemCatChem* **2018**, *10*, 5194–5200. [CrossRef]
6. Qu, K.; Zheng, Y.; Dai, S.; Qiao, S.Z. Graphene oxide-polydopamine derived N, S-codoped carbon nanosheets as superior bifunctional electrocatalysts for oxygen reduction and evolution. *Nano Energy* **2016**, *19*, 373–381. [CrossRef]
7. Zhang, J.; Zhao, Z.; Xia, Z.; Dai, L. A metal-free bifunctional electrocatalyst for oxygen reduction and oxygen evolution reactions. *Nat Nano* **2015**, *10*, 444–452. [CrossRef]
8. Chu, Y.; Gu, L.; Ju, X.; Du, H.; Zhao, J.; Qu, K. Carbon Supported Multi-Branch Nitrogen-Containing Polymers as Oxygen Reduction Catalysts. *Catalysts* **2018**, *8*, 245. [CrossRef]
9. Zheng, Y.; Jiao, Y.; Qiao, S.Z. Engineering of Carbon-Based Electrocatalysts for Emerging Energy Conversion: From Fundamentality to Functionality. *Adv. Mater.* **2015**, *27*, 5372–5378. [CrossRef]
10. Jin, H.; Guo, C.; Liu, X.; Liu, J.; Vasileff, A.; Jiao, Y.; Zheng, Y.; Qiao, S.-Z. Emerging Two-Dimensional Nanomaterials for Electrocatalysis. *Chem. Rev.* **2018**, *118*, 6337–6408. [CrossRef]

11. Lai, L.; Potts, J.R.; Zhan, D.; Wang, L.; Poh, C.K.; Tang, C.; Gong, H.; Shen, Z.; Lin, J.; Ruoff, R.S. Exploration of the active center structure of nitrogen-doped graphene-based catalysts for oxygen reduction reaction. *Energy Environ. Sci.* **2012**, *5*, 7936–7942. [CrossRef]

12. Yang, H.B.; Miao, J.; Hung, S.-F.; Chen, J.; Tao, H.B.; Wang, X.; Zhang, L.; Chen, R.; Gao, J.; Chen, H.M.; et al. Identification of catalytic sites for oxygen reduction and oxygen evolution in N-doped graphene materials: Development of highly efficient metal-free bifunctional electrocatalyst. *Sci. Adv.* **2016**, *2*, e1501122. [CrossRef] [PubMed]

13. Lv, J.-J.; Li, Y.; Wu, S.; Fang, H.; Li, L.-L.; Song, R.-B.; Ma, J.; Zhu, J.-J. Oxygen Species on Nitrogen-Doped Carbon Nanosheets as Efficient Active Sites for Multiple Electrocatalysis. *ACS Appl. Mater. Interfaces* **2018**, *10*, 11678–11688. [CrossRef] [PubMed]

14. Liu, Z.; Zhao, Z.; Wang, Y.; Dou, S.; Yan, D.; Liu, D.; Xia, Z.; Wang, S. In Situ Exfoliated, Edge-Rich, Oxygen-Functionalized Graphene from Carbon Fibers for Oxygen Electrocatalysis. *Adv. Mater.* **2017**, *29*, 1606207. [CrossRef] [PubMed]

15. Kordek, K.; Jiang, L.; Fan, K.; Zhu, Z.; Xu, L.; Al-Mamun, M.; Dou, Y.; Chen, S.; Liu, P.; Yin, H.; et al. Two-Step Activated Carbon Cloth with Oxygen-Rich Functional Groups as a High-Performance Additive-Free Air Electrode for Flexible Zinc–Air Batteries. *Adv. Energy Mater.* **2018**, *2018*, 1802936. [CrossRef]

16. Liang, J.; Jiao, Y.; Jaroniec, M.; Qiao, S.Z. Sulfur and Nitrogen Dual-Doped Mesoporous Graphene Electrocatalyst for Oxygen Reduction with Synergistically Enhanced Performance. *Angew. Chem. Int. Ed.* **2012**, *51*, 11496–11500. [CrossRef] [PubMed]

17. Qu, K.; Zheng, Y.; Zhang, X.; Davey, K.; Dai, S.; Qiao, S.Z. Promotion of Electrocatalytic Hydrogen Evolution Reaction on Nitrogen-Doped Carbon Nanosheets with Secondary Heteroatoms. *ACS Nano* **2017**, *11*, 7293–7300. [CrossRef] [PubMed]

18. Ismagilov, Z.R.; Shalagina, A.E.; Podyacheva, O.Y.; Ischenko, A.V.; Kibis, L.S.; Boronin, A.I.; Chesalov, Y.A.; Kochubey, D.I.; Romanenko, A.I.; Anikeeva, O.B.; et al. Structure and electrical conductivity of nitrogen-doped carbon nanofibers. *Carbon* **2009**, *47*, 1922–1929. [CrossRef]

19. Qu, K.; Zheng, Y.; Jiao, Y.; Zhang, X.; Dai, S.; Qiao, S.Z. Polydopamine-Inspired, Dual Heteroatom-Doped Carbon Nanotubes for Highly Efficient Overall Water Splitting. *Adv. Energy Mater.* **2017**, *7*, 1602068. [CrossRef]

20. Qu, K.; Wang, Y.; Zhang, X.; Chen, H.; Li, H.; Chen, B.; Zhou, H.; Li, D.; Zheng, Y.; Dai, S. Polydopamine-Derived, In Situ N-Doped 3D Mesoporous Carbons for Highly Efficient Oxygen Reduction. *ChemNanoMat* **2018**, *4*, 417–422. [CrossRef]

21. Chen, S.; Duan, J.; Jaroniec, M.; Qiao, S.-Z. Nitrogen and Oxygen Dual-Doped Carbon Hydrogel Film as a Substrate-Free Electrode for Highly Efficient Oxygen Evolution Reaction. *Adv. Mater.* **2014**, *26*, 2925–2930. [CrossRef] [PubMed]

22. Qu, K.; Zheng, Y.; Dai, S.; Qiao, S.Z. Polydopamine-graphene oxide derived mesoporous carbon nanosheets for enhanced oxygen reduction. *Nanoscale* **2015**, *7*, 12598–12605. [CrossRef] [PubMed]

23. Datsyuk, V.; Kalyva, M.; Papagelis, K.; Parthenios, J.; Tasis, D.; Siokou, A.; Kallitsis, I.; Galiotis, C. Chemical oxidation of multiwalled carbon nanotubes. *Carbon* **2008**, *46*, 833–840. [CrossRef]

24. Jiang, Y.; Yang, L.; Sun, T.; Zhao, J.; Lyu, Z.; Zhuo, O.; Wang, X.; Wu, Q.; Ma, J.; Hu, Z. Significant contribution of intrinsic carbon defects to oxygen reduction activity. *ACS Catal.* **2015**, *5*, 6707–6712. [CrossRef]

25. Ishizaki, T.; Chiba, S.; Kaneko, Y.; Panomsuwan, G. Electrocatalytic activity for the oxygen reduction reaction of oxygen-containing nanocarbon synthesized by solution plasma. *J. Mater. Chem. A* **2014**, *2*, 10589–10598. [CrossRef]

26. Lu, Z.; Chen, G.; Siahrostami, S.; Chen, Z.; Liu, K.; Xie, J.; Liao, L.; Wu, T.; Lin, D.; Liu, Y. High-efficiency oxygen reduction to hydrogen peroxide catalysed by oxidized carbon materials. *Nat. Catal.* **2018**, *1*, 156. [CrossRef]

27. Zhang, R.; Jing, X.; Chu, Y.; Wang, L.; Kang, W.; Wei, D.; Li, H.; Xiong, S. Nitrogen/oxygen co-doped monolithic carbon electrodes derived from melamine foam for high-performance supercapacitors. *J. Mater. Chem. A* **2018**, *6*, 17730–17739. [CrossRef]

28. Zhou, R.; Zheng, Y.; Jaroniec, M.; Qiao, S.-Z. Determination of the Electron Transfer Number for the Oxygen Reduction Reaction: From Theory to Experiment. *ACS Catal.* **2016**, *6*, 4720–4728. [CrossRef]

29. Tian, G.-L.; Zhang, Q.; Zhang, B.; Jin, Y.-G.; Huang, J.-Q.; Su, D.S.; Wei, F. Toward Full Exposure of "Active Sites": Nanocarbon Electrocatalyst with Surface Enriched Nitrogen for Superior Oxygen Reduction and Evolution Reactivity. *Adv. Funct. Mater.* **2014**, *24*, 5956–5961. [CrossRef]

30. Gao, M.-R.; Xu, Y.-F.; Jiang, J.; Zheng, Y.-R.; Yu, S.-H. Water Oxidation Electrocatalyzed by an Efficient Mn3O4/CoSe2 Nanocomposite. *J. Am. Chem. Soc.* **2012**, *134*, 2930–2933. [CrossRef]

31. Liang, Y.; Li, Y.; Wang, H.; Zhou, J.; Wang, J.; Regier, T.; Dai, H. Co$_3$O$_4$ nanocrystals on graphene as a synergistic catalyst for oxygen reduction reaction. *Nat. Mater.* **2011**, *10*, 780–786. [CrossRef] [PubMed]

32. Masa, J.; Xia, W.; Sinev, I.; Zhao, A.; Sun, Z.; Grützke, S.; Weide, P.; Muhler, M.; Schuhmann, W. MnxOy/NC and CoxOy/NC Nanoparticles Embedded in a Nitrogen-Doped Carbon Matrix for High-Performance Bifunctional Oxygen Electrodes. *Angew. Chem. Int. Ed.* **2014**, *53*, 8508–8512. [CrossRef] [PubMed]

33. Ma, T.Y.; Dai, S.; Jaroniec, M.; Qiao, S.Z. Graphitic Carbon Nitride Nanosheet–Carbon Nanotube Three-Dimensional Porous Composites as High-Performance Oxygen Evolution Electrocatalysts. *Angew. Chem. Int. Ed.* **2014**, *53*, 7281–7285. [CrossRef] [PubMed]

34. Ma, T.Y.; Dai, S.; Jaroniec, M.; Qiao, S.Z. Metal–Organic Framework Derived Hybrid Co3O4-Carbon Porous Nanowire Arrays as Reversible Oxygen Evolution Electrodes. *J. Am. Chem. Soc.* **2014**, *136*, 13925–13931. [CrossRef] [PubMed]

35. Mao, S.; Wen, Z.; Huang, T.; Hou, Y.; Chen, J. High-performance bi-functional electrocatalysts of 3D crumpled graphene-cobalt oxide nanohybrids for oxygen reduction and evolution reactions. *Energy Environ. Sci.* **2014**, *7*, 609–616. [CrossRef]

36. Ma, T.Y.; Ran, J.; Dai, S.; Jaroniec, M.; Qiao, S.Z. Phosphorus-Doped Graphitic Carbon Nitrides Grown In Situ on Carbon-Fiber Paper: Flexible and Reversible Oxygen Electrodes. *Angew. Chem. Int. Ed.* **2015**, *54*, 4646–4650. [CrossRef] [PubMed]

37. Tian, G.-L.; Zhao, M.-Q.; Yu, D.; Kong, X.-Y.; Huang, J.-Q.; Zhang, Q.; Wei, F. Nitrogen-Doped Graphene/Carbon Nanotube Hybrids: In Situ Formation on Bifunctional Catalysts and Their Superior Electrocatalytic Activity for Oxygen Evolution/Reduction Reaction. *Small* **2014**, *10*, 2251–2259. [CrossRef]

38. Gorlin, Y.; Jaramillo, T.F. A Bifunctional Nonprecious Metal Catalyst for Oxygen Reduction and Water Oxidation. *J. Am. Chem. Soc.* **2010**, *132*, 13612–13614. [CrossRef]

39. Liu, Q.; Jin, J.; Zhang, J. NiCo2S4@graphene as a Bifunctional Electrocatalyst for Oxygen Reduction and Evolution Reactions. *ACS Appl. Mater. Interfaces* **2013**, *5*, 5002–5008. [CrossRef]

40. Lee, D.U.; Kim, B.J.; Chen, Z. One-pot synthesis of a mesoporous NiCo2O4 nanoplatelet and graphene hybrid and its oxygen reduction and evolution activities as an efficient bi-functional electrocatalyst. *J. Mater. Chem. A* **2013**, *1*, 4754–4762. [CrossRef]

41. Su, Y.; Zhu, Y.; Jiang, H.; Shen, J.; Yang, X.; Zou, W.; Chen, J.; Li, C. Cobalt nanoparticles embedded in N-doped carbon as an efficient bifunctional electrocatalyst for oxygen reduction and evolution reactions. *Nanoscale* **2014**, *6*, 15080–15089. [CrossRef] [PubMed]

42. Zheng, Y.; Jiao, Y.; Chen, J.; Liu, J.; Liang, J.; Du, A.; Zhang, W.; Zhu, Z.; Smith, S.C.; Jaroniec, M.; et al. Nanoporous Graphitic-C3N4@Carbon Metal-Free Electrocatalysts for Highly Efficient Oxygen Reduction. *J. Am. Chem. Soc.* **2011**, *133*, 20116–20119. [CrossRef] [PubMed]

43. Li, L.; Yang, H.; Miao, J.; Zhang, L.; Wang, H.-Y.; Zeng, Z.; Huang, W.; Dong, X.; Liu, B. Unraveling Oxygen Evolution Reaction on Carbon-Based Electrocatalysts: Effect of Oxygen Doping on Adsorption of Oxygenated Intermediates. *ACS Energy Lett.* **2017**, *2*, 294–300. [CrossRef]

 *catalysts*

Article

Facile Synthesis and Characterization of Two Dimensional SnO$_2$-Decorated Graphene Oxide as an Effective Counter Electrode in the DSSC

Mohamed S. Mahmoud [1,2], **Moaaed Motlak** [3] **and Nasser A. M. Barakat** [1,*]

1   Chemical Engineering Department, Minia University, El-Minia 61111, Egypt; msmm122@yahoo.com
2   Collage of Applied Science, Department of Engineering, Sohar 311, Oman
3   Department of Physics, Collage of Science, University of Anbar, Al Ramadi 31001, Iraq; moaaed.motlak@yahoo.com
*   Correspondence: nasbarakat@minia.edu.eg; Tel.: +20-8623234008

Received: 1 December 2018; Accepted: 22 January 2019; Published: 1 February 2019

Abstract: SnO$_2$-decorated graphene oxide (SnO$_2$/GO) was synthesized by the modified Hummers's method, followed by a chemical incorporation of SnO$_2$ nanoparticles. Then, the nanocomposite was used as anon-precious counter electrode in a dye-sensitized solar cell (DSSC). Although GO has a relatively poor electrical conductivity depending essentially on the extent of the graphite oxidation, presence of SnO$_2$ enhanced its structural and electrochemical properties. The Pt-free counter electrode exhibited a distinct catalytic activity toward iodine reduction and a low resistance to electron transfer. Moreover, the decorated GO provided extra active sites for reducing I$_3^-$ at the interface of the CE/electrolyte. In addition, the similarity of the dopant in the GO film and the fluorine-doped tin oxide (FTO) substrate promoted a strong assimilation between them. Therefore, SnO$_2$-decorated GO, as a counter electrode, revealed an enhanced photon to electron conversion efficiency of 4.57%. Consequently, the prepared SnO$_2$/GO can be sorted as an auspicious counter electrode for DSSCs.

Keywords: dye sensitized solar cell; SnO$_2$-decorated graphene oxide; counter electrode; solar energy

---

1. Introduction

Basically, the conversion of solar energy into electricity proceeds through direct or indirect routes. Currently, direct conversion of solar radiation-to-electricity is receiving the maximum attention. In this regard, silicon-based photovoltaic cell (PV panel) is the most widely used. However, PV cells are suffering from high fabrication cost ($/kW) because their function counts on the presence of a relatively thick layer of doped silicon to obtain an acceptable photon capture rate. Comparably, the dye sensitized solar cell (DSSC) provides a low-cost transfer route of photons to electrons, due to its superior advantages over the traditional PV cells; it absorbs more solar radiation per surface area than the traditional PV cells [1,2]. Recently, the efficiency of the DSSCs has been increased to 17%, which puts it in the market race with the conventional PV [3]. Mainly, the DSSC consists of a porous layer of titanium oxide nanoparticle coated with a solar sensitive organometallic dye, the counter electrode (Pt), and the working electrolyte (e.g., the redox couple I$_3^-$/I$^-$). Its mechanism of work is discussed in details elsewhere [4,5]. Numerous efforts to enhance the solar energy-to-electricity efficiency of the DSSCs have been reported. These attempts focused on optimizing the properties of the four components of the DSSC:

1.   The sensitized dye
2.   The material of the photoanode
3.   The redox couple electrolyte and

4.   The non-precious counter electrode

In this regard, there have been several attempts to develop stable solar-sensitive dyes [3,6] and electrolytes [6,7]. Similarly, developing the photoanode has been investigated by altering the type of the mesoporous film that is coated over the photo anode glass, such as TiO_2 nanofibers [8,9], ZnO and ZnO/TiO_2 nanocomposite [10,11], SnO_2 [12,13], or CdO [14]. These oxides are mixed with the organometallic dye to form the photoanode of the DSSC.

Compared to the photoanode, the counter electrode (CE) is more expensive because it is usually fabricated from precious metals (e.g., Pt), which adds an additional capital cost. Therefore, developing effective counter electrode from cheap materials can strongly enhance the DSSC's rank. Typically, the CE should assure three attributes [15]: (i) As a catalyst, it should transfer the electron to the oxidized redox couple (I^-). (ii) As a cathode, it should collect the electrons coming from the outside circuit and get them ready to be transferred to the cell. (iii) As a mirror, it should reflect the transmitted light back to the DSSC to enhance the use of photons [16]. The main characteristics of the optimal CE are summarized as follows [17]: high catalytic activity, high electrical conductivity, maximum reflectivity, low-cost, large surface area, porous nature, optimal thickness, electrochemical and mechanical stability, energy level that matches the potential of the redox couple electrolyte, and high adhesivity with the FTO [17]. As a heterogeneous catalytic reaction is taking place on the surface of the counter electrode, the reduction of the electrolyte (e.g., iodine ion) can be considered a combination of adsorption and electrochemical reaction processes. Accordingly, to exploit their high adsorption capacity, carbon nanomaterials such as nanotubes [18], graphene [19], and nanofibers [20] have been utilized as support for the counter electrode materials. Moreover, some carbonaceous nanostructures show a distinguished activity individually [21,22]. Graphene is branded by its amazing electrical and mechanical properties. Therefore, many researchers show interest in using this nanomaterial in different applications [23]. However, its high hydrophobicity negatively affects the performance in the aqueous media due to the poor contact which adds an additional resistance to the reactants transfer. On the other hand, due to possessing oxygenated groups on the surface, graphene oxide (GO) is more hydrophilic than graphene. However, the newly added active groups decrease the electrical conductivity of GO compared to graphene. The conductivity of GO depends on the C/O ratio in the sample. During the oxidation step, the sp2 carbons are removed and replaced by sp3 ones having oxygen functionalities. This process creates a band gap by pulling the bands apart. Hence, when graphene is fully oxidized, the bands are far apart, and the GO behaves as an insulator. In the middle of this fully oxidized GO and pure graphene, it behaves as a semiconductor. Partially oxidized GO shows the appearance of sp3 regions and sp2 regions [24], and the GO can conduct through these sp2 regions via Klein tunneling [24]. In addition, insertion of a transition metal oxide in the GO layers could play a crucial role in enhancing the electron conduction of the GO. SnO_2 is a crucial semiconductor that is used as a transparent conducting oxide (TCO). This TCO normally coated over the glass substrate and used for optical applications. Hitherto, SnO_2 is usually considered as an oxygen-deficient n-type semiconductor [25]. Therefore, doping GO with SnO_2 nanoparticles may enhance the conductive properties of the GO and promote the electron transfer to the redox chattel in the DSSC. Even though there are intensive studies on alternatives of counter electrode, thus far, to our best knowledge, there is no report regarding the application of SnO_2-decorated GO as counter electrode in the DSSC. In this work, we report the synthesis of SnO_2-decorated GO, its characterization, and application as a counter electrode in the DSSC.

2. Results and Discussion

Figure 1a shows the transmission electron microscope (TEM) image of the prepared material. It is noticeable that the sample composed of a sheet of GO contains randomly distributed nanoparticles. High resolution TEM (HR-TEM) image shown in Figure 1b affirms that the attached nanoparticles appear as textured crystalline mode, which indicates that these nanoparticles are SnO_2-based, highly crystalline compound. However, the distribution of these nanoparticles is random. The higher magnification image indicates that the interlayer distance is 0.26 nm and the particle size is less than 20 nm.

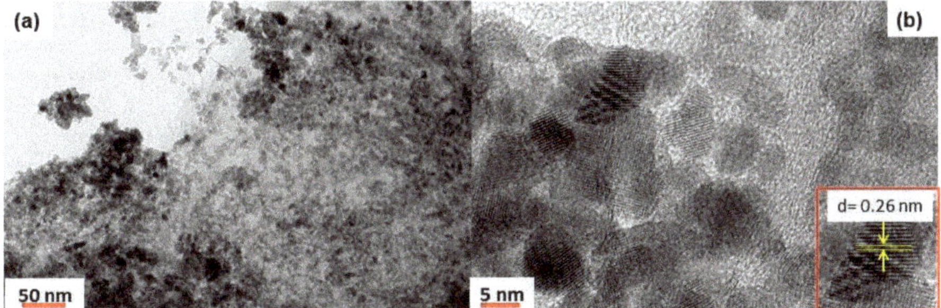

Figure 1. Transmission electron microscope (TEM) (**a**); and High resolution TEM (HR-TEM) (**b**) images of SnO$_2$-incorporated GO.

Figure 2 displays the X-ray diffraction (XRD) patterns of both pristine and decorated GO. Initially, from XRD pattern, it is easy to confirm the formation of GO from the utilized precursor. Typically, the graphite reveals a sharp peak in the XRD pattern at 2θ value of ~26°. Due to exfoliation, the graphite-identified peak disappears and another peak is formed at ~10°, while graphene is distinguished by a very broad peak centered at ~24° [26]. As shown, the GO main peak at 2θ values of 10.2°, corresponding to the (001) basal plane of GO with a d-spacing value of (d001 = 0.961 nm), appears in both samples. However, the intensity of such peak in the SnO$_2$/GO sample has a lower intensity than pure GO. Typically, the intensity of the GO main identification peak is 867 and 1957 count/s for the SnO$_2$/GO and pristine GO, respectively. The reason of low peak intensity is probably due to the heat treatment and incorporation of foreign atoms in the GO layer, which was also noticed by R. Krishna et al. [27]. In addition, the strong diffraction peaks at 2θ values of 26.38°, 33.48°, 37.76°, and 51.46° corresponding to (110), (101), (200) and (211) crystal planes (JCPDS:41-1445), respectively, confirm the incorporation of tetragonal crystals of SnO$_2$ inside the GO layer. It is also noticeable that the SnO$_2$ peaks are broadened, indicating the presence of small crystalline size. It is important to determine the crystal structure of the SnO$_2$-decorated GO due to its direct effect on the power conversion efficiency of the DSSC. According to Scherer's equation ($\tau = K\lambda/\beta\cos\theta$, where τ is the ordered (crystalline) domains mean size, λ is the wavelength of the utilized X-ray irradiation (0.1504 nm), K is a constant value with a typical value of 0.9 referring to the shape factor, β is the line broadening at half the maximum intensity (FWHM) in radians and θ is the Bragg angle), the average grain size was determined to be 5.1 nm.

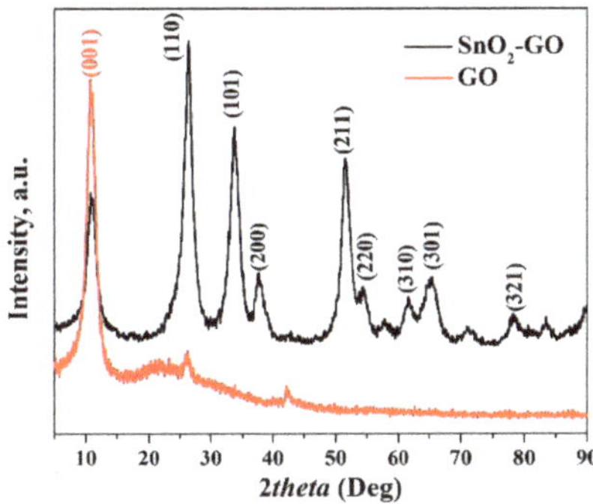

Figure 2. XRD patterns for the pristine and SnO$_2$-incorporated graphene oxide.

Figure 3 displays the Raman spectra of pristine and SnO_2-incorporated GO in the range of 100–3000 cm^{-1}. In the GO spectrum, two protuberant peaks corresponding to D and G bands, the characteristic peaks of GO, are located at 1354 and 1591 cm^{-1} with respective I_D/I_G intensities ratio of 1.07. The inset confirms the stability of the prepared GO during the achieved GO decoration process by SnO_2 nanoparticles. As shown in the inset, there is no considerable change in the I_D/I_G intensities ratio. Moreover, in the Raman spectrum, the pristine graphene reveals an additional peak (2D), indicating an increase in the average size of sp2 domains at ~2790 cm^{-1} [28]. As shown, the 2D peak does not appear in the SnO_2-GO spectrum, which confirms maintaining the composition of the GO. In other words, the utilized treatment process did not affect the content of the oxygenated groups covering the surface of GO sheets. The crystalline structure of SnO_2 is a tetragonal rutile with point group D4h [29]. Typically, three modes identifying SnO_2 can be found in the Raman spectrum: 474 cm^{-1} (E_g), 631 cm^{-1} (A_{1g}) and 775 cm^{-1} (B_{2g}) [30]. When the particle size decreases, A_{1g} and B_{2g} modes of SnO_2 are shifted to lower wave numbers and E_g mode is shifted to higher wave number [30]. The obtained results show peaks at 615, 435, 245, and 120 cm^{-1}. The band at 435 cm^{-1} probably corresponds to the (E_g) mode of the oxide. Similarly, the mode at 615 cm^{-1} can be assigned to symmetric O-Sn-O stretching (A_{1g}) shifts due to the relatively small particle size of SnO_2 nanoparticles. The position and the intensity of SnO_2 peak in the Raman spectrum depends on the crystal size. The vibrational mode (B_{1g}) peak appears only along with nanomaterials. Thus, in the Raman spectrum, the presence of a peak at 120 cm^{-1} is associated with non-degenerated B_{1g} mode of SnO_2 and its appearance is caused by lowering the surface phonon frequencies that leads to increase in the inter-atomic distance on the surface [30].

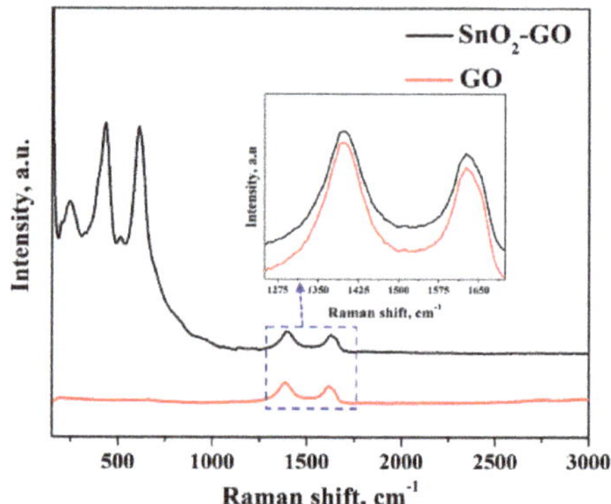

Figure 3. Raman spectroscopy analyses for the pristine and SnO_2-decorated graphene oxide. The inset displays high magnification of the D and G bands.

Overall, based on the aforementioned analytical technique, it is safe to claim that the proposed material is GO sheets decorated by small and highly crystalline SnO_2 nanoparticles.

It is worth mentioning that the function of the redox shuttle (I_3^-/I^-) is very important to ensure a stable transfer of electrons between the photoanode and the counter electrode of a DSSC. The photon-induced excitation of the dye leads to transfer of electrons to the oxide semiconductor. Then, the electron donor (I_3^-) must rapidly return the excited dye back to its ground state. Afterwards, the electron acceptor (I^-) should move to the CE to get the missing electrons to regenerate the electron donor (I_3^-). The electron migration through the external circuit will close the circuit [2,4].

To verify the electrocatalytic activity of the as-prepared SnO_2-incorporated GO, cyclic voltammetry (CV) analysis of the prepared catalyst was achieved using a conventional three-electrode system. SnO_2-doped GO, Pt wire and Ag/AgCl served as working, counter and reference electrodes, respectively. The electrolyte solution consisted of 10 mM LiI, 1 mM I_2, and 0.1 M $LiClO_4$ in acetonitrile. Figure 4a shows the cyclic voltammograms (CV) at scan rate of 25 mV/s. It is clear that the introduced SnO_2/GO shows both oxidation and reduction peaks, which indicate achieving the following reactions [20].

$$I_3^- + 2e^- \rightarrow 3I^- \tag{1}$$

$$3I_2 + 2e^- \rightarrow 2I_3^- \tag{2}$$

The counter electrode exhibits high redox current densities, indicating that the speed of the redox reaction on the surface of the modified GO is fast. Besides the activity of the SnO_2 nanoparticles attached on the surface of the GO, the enhancement in the surface area after the decoration process can be assigned as a second reason for the observed high performance. Typically, the measured surface area of the pristine and decorated GO was 159.7 ± 2 and 210.5 ± 3 m^2/g, respectively. The increase in the surface area can be attributed to the high exfoliation rate upon addition of the tin precursor. Therefore, the improved specific surface area of the modified GO can be considered as a stimulator for such boost of the electrocatalytic activity toward the reduction of I_3^- ions. Incidentally, the modified GO is responsible for the efficient reduction of I_3^- ions by providing sufficient superficial area and promoting fast electronic transformation, which results in higher J_{sc} and solar-electrical conversion efficiency in the DSSC.

Figure 4b illustrates CV curves of the SnO_2-incorporated GO electrode in the prepared electrolyte solution at different scan rates (10–50 mV/s). It is clear that altering the scan rate increased the current density values. In addition, it slightly shifted the potential peaks of the anodic and cathodic reactions. The inset of Figure 4b shows the influence of the scan rate on the anodic and cathodic peak currents. It is noticeable that they are linearly dependent on the scan rate. Such finding discloses that the rate of diffusion of I_3^- ion is the hindering step in the oxidation–reduction process [20].

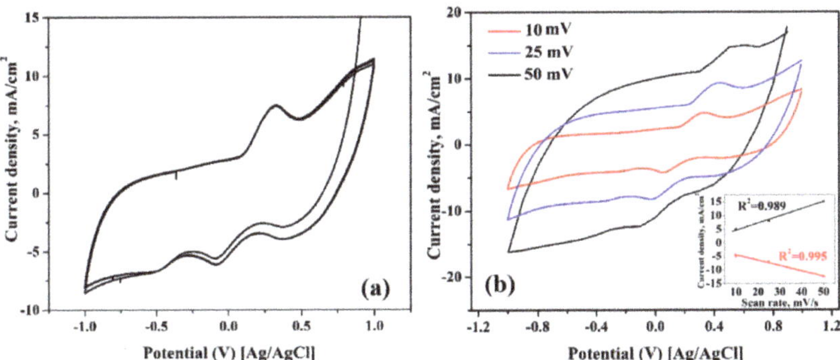

Figure 4. (a) Cyclic voltammetry analyses of the introduced SnO_2/GO in presence of iodine solution with a scan rate of 25 mV/s in the potential range from -1 to 1 V vs. Ag/AgCl; and (b) the cyclic voltammograms at different scan rates.

Performing the electrochemical impedance spectroscopy (EIS) analysis of a DSSC helps to get insight about some important performance parameters, such as charge transport through the photoanode and near the counter electrode, charge transfer due to electron back reaction, and capacitive accumulation of the charges at different processes in the cell. It was used to measure the resistances of the individual interfaces of the fabricated DSSC based on the introduced SnO_2/GO counter electrode. The merit of EIS spectrum over the I–V curve is that EIS indicates the response of the solar-induced current on the electrical bias voltage values, which cannot be indicated through the I–V curve.

The impedance spectrum of a DSSC typically shows three semicircles in the Nyquist plot. According to the direction of decreasing frequency, the first semicircle corresponds to the charge transfer processes which happens in the CE/electrolyte interface with a characteristic frequency ω_{CE}; the second or middle semicircle corresponds to the diffusion of the electron through the TiO_2 layer and the reversible electron reaction with the oxidized state of the redox species at the TiO_2/electrolyte boundary; and the third semicircle at the low frequency region corresponds to the diffusion of I_3^- in the electrolyte, with a characteristic frequency ω_D [31]. The characteristic frequency for electron transport or diffusion (ω_d) appears in the middle semicircle at the high frequency region, while the peak frequency (ω_k) of that semicircle corresponds to the electron reversible reaction. Figure 5 shows the Nyquist and Bode plots of GO and SnO_2/GO as counter electrodes in the DSSC. In the case of GO, as shown in Figure 5A, only one semicircle appears, indicating an insignificant charge transfer at the GO/electrolyte interface. For the SnO_2/GO plot, two semicircles signify two different kinds of impedance. In the high-frequency region (10^3–10^6 Hz), the semicircle denotes the resistance due to the charge transfer (R_{CE}) at the counter electrode/electrolyte interface. In the middle-frequency (1–10^3 Hz) region, the semicircle denotes the diffusion through the TiO_2 layer and electron reversible reaction with the oxidized state of the redox species at the TiO_2/electrolyte boundary (R_{CT}). However, in the low-frequency region (0.1–1 Hz), no semicircle of the complete response appears, which could be attributed to the small distance between the working and counter electrodes and the low viscosity of the electrolyte.

These data indicate that the low (R_{CE}) is related to a comparatively high rate of electron transfer at the interface of electrolyte/counter electrode, which leads to improve the catalytic activity toward the redox couple in the electrolyte. In the DSSC cell, the (R_{CE}) value was found to be 43.1 Ω, which indicates that the SnO_2/GO revealed an effective catalytic activity toward the reduction of I_3^- ions in redox electrolyte. Moreover, the cell shows low series resistance (Rs) value of 23.6 Ω, which is attributed to an excellent conductivity of SnO_2/GO film and strong contact between the film and FTO substrate [20].

The catalytic activity of the SnO_2-incorporated GO counter electrode can be presented using current density (J), which is calculated from the charge transfer resistance (R_{CT}) [32]:

$$R_{CT} = RT/nFJ \tag{3}$$

where R is the gas constant (J/mol·K), T is temperature (K), n is the number of involved electrons ($n = 2$), and F is the Faraday constant.

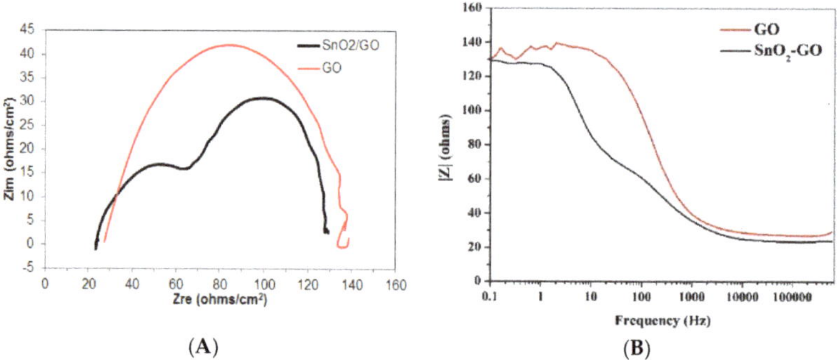

Figure 5. Electrochemical impedance of the DSSC fabricated with GO and SnO_2-incorporated GO-based counter electrode: (**A**) Nyquist plot; and (**B**) Bode diagram.

Figure 6 depicts the J–V curves of DSSCs under metal halide lamp illumination. Table 1 displays the photovoltaic parameters from J–V curves and a brief comparison with other counter electrodes. The data indicate that the cell based on the introduced SnO_2/GO exhibited an energy conversion efficiency of 4.57%, which is slightly lower than the standard Pt counter electrode. Such energy conversion efficiency of the cell is linked to the activity of the counter electrode, which synergizes the reduction reaction at the interface of electrolyte/counter electrode, which in turn assists the regeneration of the ground state of the organometallic dye at the interface of electrolyte/photoanode. As shown in Table 1, the other parameters for the fabricated cell are V_{oc} of 0.67 V, J_{sc} of 14.04 mA^{-2}

cm and FF of 0.49. These results are lower than those reported from GO–Pt composite fabricated by low-temperature electrodeposition process, which revealed an energy conversion efficiency of 7.05% [33]. However, the authors used Pt precious metal, which is not aligned with our goals.

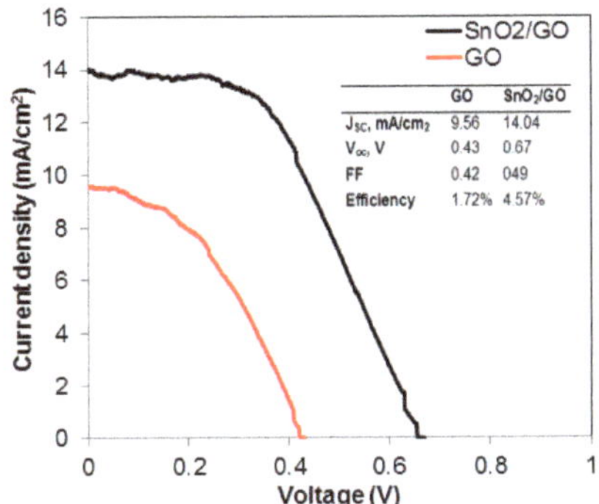

Figure 6. Photovoltaic characteristics (J–V) of fabricated DSSCS using SnO_2-incorporated GO based counter electrode measured under the irradiance of AM 1.5 G sunlight of 100 mW cm^{-2}.

Table 1. The detailed photovoltaic parameters from J–V curves and a brief comparison with other works.

Type of CE	η, %	FF	V_{OC}, V	J_{sc}, mA/cm^2	Ref.
Pt layer	5.9	0.51	0.718	16.12	[20]
CoS/rGO	9.39	0.63	0.764	19.42	[34]
Mw-Pt NPs@GO	7.96	0.71	0.7	16.02	[35]
GO/FTO	3.99	0.38	0.71	14.02	[36]
GO-ED Pt/FTO	7.05	0.66	0.71	14.98	[37]
POMA-FGO (1%)	7.26	0.61	0.73	16.31	[38]
CuS5/FTO5 µm	1.12	0.52	0.37	5.81	[39]
SnO_2/GO	4.57	0.49	0.67	14.04	This study

3. Materials and Methods

3.1. Catalyst Preparation

To prepare the electrode, all the involved chemicals, e.g., $SnCl_2$, thiourea, graphite, and $KMnO_4$, were obtained from Sigma Aldrich Co., Seoul, South Korea. The procedure started with slowly dissolving 0.42 g of tin chloride in 20 mL distilled water. Parallelly, 0.15 g of thiourea were also dissolved in 20 mL distilled water individually for 20 min. Then, the two solutions were mixed using a magnetic stirrer for 1 h. The GO was prepared by a modified Hummers's method. The next step was to disperse 0.2 g of GO in 100 mL distilled water, and then ultra-sonication for 40 min. Afterwards, the two mixtures were mixed for 20 min, and then 200 mL of hydrazine were added just before the hydrothermal step. Next, hydrothermal treatment was done for 10 h at 180 °C. Finally, the sample was filtered and dried at 60 °C overnight.

3.2. The DSSC Fabrication

The fabrication of DSSCs consisted of three steps [18,39]: fabrication of the photoanode, fabrication of the counter electrode and the addition of the working electrolyte. Typically, FTO substrate glass (FTO 10 Ω/sq.) was used to prepare the photoanode. Preparation started by plating nanocrystalline TiO_2 (Degussa P-25) thin film over FTO glass by simple doctor blade technique. The prepared photoanode had an active area of 0.25 cm^2 with the film thickness of 8–10 µm. Then, the FTO was annealed at 450 °C for 30 min. Afterwards, the photoanodes were immersed in the dye solution consisting of

0.3 mM ruthenium(II) 535 bis-TBA (N-719, Solaronix; Sigma, Seoul, South Korea), rinsed with ethanol and dried for 24 h under nitrogen flow. Similarly, the SnO_2-decorated GO was pasted over the FTO glass by simple doctor blade technique. The obtained film was cleaned by ethanol and then dried at 60 °C for 30 min. To complete the fabrication of DSSC, the SnO_2-decorated GO counter electrode was placed over the dye-adsorbed TiO_2 photoanode and sealing was made using Surlyn sheet (SX 1170-60, Solaronix, 60 μm thick) between the two electrodes. Finally, the electrolyte, consisting of lithium iodide (LiI 0.5 M), Iodine (I_2 0.05 mM), and tetra-butyl-pyridine ($C_9H_{13}N$ 0.2 M) in acetonitrile, was inserted into the holes in the counter electrode using a syringe.

3.3. Characterization and Application

The TEM images of the prepared catalyst were retrieved using JEOL JEM 2010 transmission electron microscope, working at a voltage of 200 kV, (JEOL Ltd., Tokyo, Japan). Clarification of the distance of the TEM images was acquired by ImageJ 1.47v software. To determine the crystal structure of the prepared catalyst, Rigaku X-ray diffractometer (XRD, Rigaku, Tokyo, Japan) with Cu Kα (λ = 1.5406 Å) was utilized. The Raman spectrum was also acquired using Dispersive Raman spectrometer (BRUKER-SENTERRA, Boerdestr, Warstein, Germany) equipped with an integral microscope (Olympus, Tokyo, Japan). The excitation source was neodymium-doped yttrium aluminum garnet (Nd/YAG) laser (532 nm), and providing a power of 20 mW on the sample, and analyzed by X-ray Diffraction Microanalysis (XPERT-PRO-P-Analysis, The Woodlands, TX, USA). The electrochemical impedance spectroscopy (EIS) and cyclic Voltammetry (CV) measurements were performed using a VersaSTAT 4 (AMETEK, New York, NY, USA) electrochemical analyzer using a conventional three-electrode cell structure. In the utilized cell, glassy carbon, Pt and Ag/AgCl served as working, counter and reference electrodes, respectively. To deposit the proposed GO-based material on the active surface of the working electrode, 2 mg of the function material were added to 20 μL Nafion solution (5 wt %) and 400 μL isopropanol. Afterwards, the suspension was subjected to utltra-sonication process for 20 min. Then, 15 μL of the suspension were poured in three steps on the electrode active area. Finally, the electrode was dried at 80 °C. Current–voltage characteristics of DSSCs were measured by using digital Multimeters (Model 2000, Keithley, Filderstadt, Germany) and a variable load. As a simulation of the solar radiation, a 1000 W metal halide lamp served as a light source, and its light intensity (or radiant power) was adjusted to simulate mornign. Then, 1.5 radiation at 100 mW cm^{-2} with a Si photo detector fitted with a KG-5 filter (Schott, Cheney, KS, USA) as a reference was calibrated at NREL (New York, NY, USA). The metal uptake, during the decoration process, was determined by estimating the remaining tin ions in the initial solution using the titration procedure. Briefly, the solution was hot treated with hydrochloric acid, and then drops of starch indicator solution were added. Finally, the solution was titrated against iodate–iodide standard solution. The uptake was determined to be around 15 wt %.

4. Conclusions

SnO_2 nanoparticles-decorated graphene oxide can be synthesized by hydrothermal treatment of GO prepared by Hummers's method in presence of tin chloride. The active chemical groups covering the prepared GO upon using the chemical route lead to a homogeneous distribution for the metallic nanoparticles. Moreover, these groups procure to a good attachment of the inorganic oxide with the carbonaceous support. The prepared SnO_2/GO catalyst improves the electrocatalytic activity toward iodine redox reaction. Moreover, a decrease in the electron transfer was observed due to the harmony between the metallic nanoparticles in the proposed composite and the tin oxide existing in the utilized transparent glass electrodes (FTO). Accordingly, exploiting the prepared catalyst as a counter electrode in the DSSC showed a relatively good energy conversion with an efficiency of 4.57%. Overall, it can be claimed that the SnO_2-decorated GO might be a new effective counter electrode for DSSCs.

Author Contributions: M.S.M. helped in writing the original draft of the manuscript. M.M. helped in the formal analysis. N.A.M.B. helped in supervision and project administration.

Acknowledgments: The authors deeply thank M. Shaheer Akhtar (Chonbuk National University) for measuring the J–V data.

Conflicts of Interest: The authors declare no conflict of interest.

References

1. Mathew, S.; Yella, A.; Gao, P.; Humphry-Baker, R.; Curchod, B.F.; Ashari-Astani, N.; Tavernelli, I.; Rothlisberger, U.; Nazeeruddin, M.K.; Grätzel, M. Dye-sensitized solar cells with 13% efficiency achieved through the molecular engineering of porphyrin sensitizers. *Nat. Chem.* **2014**, *6*, 242. [CrossRef] [PubMed]

2. Ning, Z.; Fu, Y.; Tian, H. Improvement of dye-sensitized solar cells: What we know and what we need to know. *Energy Environ. Sci.* **2010**, *3*, 1170–1181. [CrossRef]

3. Lee, C.-P.; Li, C.-T.; Ho, K.-C. Use of organic materials in dye-sensitized solar cells. *Mater. Today* **2017**, *20*, 267–283. [CrossRef]

4. Gong, J.; Sumathy, K.; Qiao, Q.; Zhou, Z. Review on dye-sensitized solar cells (DSSCs): Advanced techniques and research trends. *Renew. Sustain. Energy Rev.* **2017**, *68*, 234–246. [CrossRef]

5. Sharma, S.; Siwach, B.; Ghoshal, S.; Mohan, D. Dye sensitized solar cells: From genesis to recent drifts. *Renew. Sustain. Energy. Rev.* **2017**, *70*, 529–537. [CrossRef]

6. Ye, M.; Wen, X.; Wang, M.; Iocozzia, J.; Zhang, N.; Lin, C.; Lin, Z. Recent advances in dye-sensitized solar cells: From photoanodes, sensitizers and electrolytes to counter electrodes. *Mater. Today* **2015**, *18*, 155–162. [CrossRef]

7. Wu, J.; Lan, Z.; Lin, J.; Huang, M.; Huang, Y.; Fan, L.; Luo, G. Electrolytes in dye-sensitized solar cells. *Chem. Rev.* **2015**, *115*, 2136–2173. [CrossRef]

8. Mahmoud, M.S.; Akhtar, M.S.; Mohamed, I.M.; Hamdan, R.; Dakka, Y.A.; Barakat, N.A. Demonstrated photons to electron activity of S-doped TiO_2 nanofibers as photoanode in the DSSC. *Mater. Lett.* **2018**, *225*, 77–81. [CrossRef]

9. Joshi, P.; Zhang, L.; Davoux, D.; Zhu, Z.; Galipeau, D.; Fong, H.; Qiao, Q. Composite of TiO_2 nanofibers and nanoparticles for dye-sensitized solar cells with significantly improved efficiency. *Energy Environ. Sci.* **2010**, *3*, 1507–1510. [CrossRef]

10. Manthina, V.; Baena, J.P.C.; Liu, G.; Agrios, A.G. ZnO–TiO_2 Nanocomposite Films for High Light Harvesting Efficiency and Fast Electron Transport in Dye-Sensitized Solar Cells. *J. Phys. Chem. C* **2012**, *116*, 23864–23870. [CrossRef]

11. Yang, M.; Dong, B.; Yang, X.; Xiang, W.; Ye, Z.; Wang, E.; Wan, L.; Zhao, L.; Wang, S. TiO_2 nanoparticle/nanofiber—ZnO photoanode for the enhancement of the efficiency of dye-sensitized solar cells. *RSC Adv.* **2017**, *7*, 41738–41744. [CrossRef]

12. Basu, K.; Benetti, D.; Zhao, H.; Jin, L.; Vetrone, F.; Vomiero, A.; Rosei, F. Enhanced photovoltaic properties in dye sensitized solar cells by surface treatment of SnO_2 photoanodes. *Sci. Rep.* **2016**, *6*, 23312. [CrossRef] [PubMed]

13. Zhang, Y.; Qi, T.; Wang, Q.; Zhang, Y.; Wang, D.; Zheng, W. Preparation of SnO_2/rGO Photoanode and Its Effect on the Property of Dye-Sensitized Solar Cells. *IEEE J. Photovolt.* **2017**, *7*, 399–403. [CrossRef]

14. Bykkam, S.; Kalagadda, B.; Kalagadda, V.R.; Ahmadipour, M.; Chakra, C.S.; Rajendar, V. Effect of Few-Layered Graphene-Based CdO Nanocomposite-Enhanced Power Conversion Efficiency of Dye-Sensitized Solar Cell. *J. Electron. Mater.* **2018**, *47*, 620–626. [CrossRef]

15. Thomas, S.; Deepak, T.; Anjusree, G.; Arun, T.; Nair, S.V.; Nair, A.S. A review on counter electrode materials in dye-sensitized solar cells. *J. Mater. Chem. A* **2014**, *2*, 4474–4490. [CrossRef]

16. Wu, J.; Li, Y.; Tang, Q.; Yue, G.; Lin, J.; Huang, M.; Meng, L. Bifacial dye-sensitized solar cells: A strategy to enhance overall efficiency based on transparent polyaniline electrode. *Sci. Rep.* **2014**, *4*, 4028. [CrossRef] [PubMed]

17. Wu, J.; Lan, Z.; Lin, J.; Huang, M.; Huang, Y.; Fan, L.; Luo, G.; Lin, Y.; Xie, Y.; Wei, Y. Counter electrodes in dye-sensitized solar cells. *Chem. Soc. Rev.* **2017**, *46*, 5975–6023. [CrossRef] [PubMed]

18. Prasetio, A.; Subagio, A.; Purwanto, A.; Widiyandari, H. Dye-sensitized solar cell based carbon nanotube as counter electrode. *AIP Conf. Proc.* **2016**, *1710*, 030054.

19. Hung, K.-H.; Li, Y.-S.; Wang, H.-W. Dye-sensitized solar cells using graphene-based counter electrode. In Proceedings of the 12th IEEE Conference on Nanotechnology (IEEE-NANO), Birmingham, UK, 20–23 August 2012; pp. 1–12.

20. Motlak, M.; Barakat, N.A.; Akhtar, M.S.; Hamza, A.; Kim, B.-S.; Kim, C.S.; Khalil, K.A.; Almajid, A.A. High performance of NiCo nanoparticles-doped carbon nanofibers as counter electrode for dye-sensitized solar cells. *Electrochim. Acta* **2015**, *160*, 1–6. [CrossRef]

21. Lee, W.J.; Ramasamy, E.; Lee, D.Y.; Song, J.S. Efficient dye-sensitized solar cells with catalytic multiwall carbon nanotube counter electrodes. *ACS Appl. Mater. Int.* **2009**, *1*, 1145–1149. [CrossRef]

22. Wang, X.; Zhi, L.; Müllen, K. Transparent, conductive graphene electrodes for dye-sensitized solar cells. *Nano Lett.* **2008**, *8*, 323–327. [CrossRef] [PubMed]

23. Marcaccio, M.; Paolucci, F. *Making and Exploiting Fullerenes, Graphene, and Carbon Nanotubes*; Springer: Berlin, Germany, 2014.

24. Gao, W. The chemistry of graphene oxide. In *Graphene Oxide*; Springer: Berlin, Germany, 2015; pp. 61–95.

25. Rahal, A.; Benramache, S.; Benhaoua, B. Preparation of n-type semiconductor SnO_2 thin films. *J. Semicond.* **2013**, *34*, 083002. [CrossRef]

26. El-Deen, A.G.; Barakat, N.A.; Khalil, K.A.; Kim, H.Y. Development of multi-channel carbon nanofibers as effective electrosorptive electrodes for a capacitive deionization process. *J. Mater. Chem. A* **2013**, *1*, 11001–11010. [CrossRef]

27. Krishna, R.; Titus, E.; Okhay, O.; Gil, J.C.; Ventura, J.; Ramana, E.V.; Gracio, J.J. Rapid electrochemical synthesis of hydrogenated graphene oxide using Ni nanoparticles. *Int. J. Electrochem. Sci.* **2014**, *9*, 69.

28. Wei, A.; Wang, J.; Long, Q.; Liu, X.; Li, X.; Dong, X.; Huang, W. Synthesis of high-performance graphene nanosheets by thermal reduction of graphene oxide. *Mater. Res. Bull.* **2011**, *46*, 2131–2134. [CrossRef]

29. Katiyar, R.; Dawson, P.; Hargreave, M.; Wilkinson, G. Dynamics of the rutile structure. III. Lattice dynamics, infrared and Raman spectra of SnO_2. *J. Phys. C Solid Stat. Phys.* **1971**, *4*, 2421. [CrossRef]

30. Li, Z.; Shen, W.; Zhang, X.; Fang, L.; Zu, X. Controllable growth of SnO_2 nanoparticles by citric acid assisted hydrothermal process. *Colloid Surf. Physicochem. Eng. Asp.* **2008**, *327*, 17–20. [CrossRef]

31. Sarker, S.; Ahammad, A.J.S.; Seo, H.W.; Kim, D.M. Electrochemical Impedance Spectra of Dye-Sensitized Solar Cells: Fundamels and Spreadsheet Calculation. *Int. J. Photoenergy* **2014**, *2014*, 17. [CrossRef]

32. Murakami, T.N.; Grätzel, M. Counter electrodes for DSC: Application of functional materials as catalysts. *Inorg. Chim. Acta* **2008**, *361*, 572–580. [CrossRef]

33. Yu, Y.-H.; Teng, I.J.; Hsu, Y.-C.; Huang, W.-C.; Shih, C.-J.; Tsai, C.-H. Covalent bond–grafted soluble poly(o-methoxyaniline)-graphene oxide composite materials fabricated as counter electrodes of dye-sensitised solar cells. *Org. Electron.* **2017**, *42*, 209–220. [CrossRef]

34. Huo, J.; Wu, J.; Zheng, M.; Tu, Y.; Lan, Z. High performance sponge-like cobalt sulfide/reduced graphene oxide hybrid counter electrode for dye-sensitized solar cells. *J. Power Sources* **2015**, *293*, 570–576. [CrossRef]

35. Demir, E.; Savk, A.; Sen, B.; Sen, F. A novel monodisperse metal nanoparticles anchored graphene oxide as Counter Electrode for Dye-Sensitized Solar Cells. *Nano-Struct Nano-Obj.* **2017**, *12*, 41–45. [CrossRef]

36. Jang, H.-S.; Yun, J.-M.; Kim, D.-Y.; Na, S.-I.; Kim, S.-S. Transparent graphene oxide–Pt composite counter electrode fabricated by pulse current electrodeposition-for dye-sensitized solar cells. *Surf. Coat. Technol.* **2014**, *242*, 8–13. [CrossRef]

37. Satoshi, K.; Daiki, A.; Etsuo, S.; Masahiro, M. Copper Sulfide Catalyzed Porous Fluorine-Doped Tin Oxide Counter Electrode for Quantum Dot-Sensitized Solar Cells with High Fill Factor. *Int. J. Photoenergy* **2017**, *2017*, 5461030.

38. Zaaba, N.; Foo, K.; Hashim, U.; Tan, S.; Liu, W.-W.; Voon, C. Synthesis of graphene oxide using modified hummers method: Solvent influence. *Procedia Eng.* **2017**, *184*, 469–477. [CrossRef]

39. Rao, S.S.; Gopi, C.V.; Kim, S.-K.; Son, M.-K.; Jeong, M.-S.; Savariraj, A.D.; Prabakar, K.; Kim, H.-J. Cobalt sulfide thin film as an efficient counter electrode for dye-sensitized solar cells. *Electrochim. Acta* **2014**, *133*, 174–179.

 *catalysts*

Article

Ternary N, S, and P-Doped Hollow Carbon Spheres Derived from Polyphosphazene as Pd Supports for Ethanol Oxidation Reaction

Ke Yu [1], Yan Lin [1], Jinchen Fan [1,2,3,*], Qiaoxia Li [1,3], Penghui Shi [1,3], Qunjie Xu [1,3,*] and Yulin Min [1,3]

[1] Shanghai Key Laboratory of Materials Protection and Advanced Materials in Electric Power, College of Environmental and Chemical Engineering, Shanghai University of Electric Power, Shanghai 200090, China; pjyuke1994@gmail.com (K.Y.); yanlinshiep2019@gmail.com (Y.L.); liqiaoxia2003@126.com (Q.L.); shipenghui@shiep.edu.cn (P.S.); minyulin@shiep.edu.cn (Y.M.)
[2] Department of Chemical Engineering and Biointerfaces Institute, University of Michigan, Ann Arbor, MI 48109, USA
[3] Shanghai Institute of Pollution Control and Ecological Security, Shanghai 200092, China
* Correspondence: jinchen.fan@shiep.edu.cn (J.F.); xuqunjie@shiep.edu.cn (Q.X.)

Received: 6 January 2019; Accepted: 22 January 2019; Published: 26 January 2019

Abstract: Ethanol oxidation reaction (EOR) is an important electrode reaction in ethanol fuel cells. However, there are many problems with commercial ethanol oxidation electrocatalysts today, such as poor durability, poor anti-CO poisoning ability, and low selectivity for C–C bond cleavage. Therefore, it is very meaningful to develop a high-performance EOR catalyst. Herein, we designed ternary N, S, and P-doped hollow carbon spheres (C–N,P,S) from polyphosphazene (PCCP) as Pd supports for EOR. Using SiO_2 spheres as the templates, the PCCP was first coated on the surfaces of SiO_2 spheres by in situ polymerization. Through high-temperature pyrolysis and hydrofluoric acid-etching, the hollow PCCP has a large surface area and porous structure. After loading Pd nanoparticles (NPs), the Pd/C–N, P, S catalysts with Pd NPs decorated on the surfaces of C–N, P, S can achieve a high mass peak current density of 1686 mA mg_{Pd}^{-1}, which was 2.8 times greater than that of Pd/C. Meanwhile, the Pd/C–N, P, S catalyst also shows a better stability than that of Pd/C after a durability test of 3600s.

Keywords: ethanol oxidation reaction; palladium; hollow carbon sphere; alkaline medium

1. Introduction

Recently, direct ethanol fuel cells (DEFCs) have become the ideal solution to the energy crisis and environmental issues and, therefore, have attracted enormous interest. Specifically, DEFCs have many advantages such as high efficiency, high energy density, low pollution, and excellent electrochemical stability. Besides, the biomass fuel ethanol is widely used because of its many advantages, such as easy to store, non-toxic, simple to synthesize [1–4]. The ethanol oxidation reaction (EOR) is a crucial reaction in DEFCs, which is the most important part of determining the performance of the entire fuel cell. Platinum (Pt) is the most widely used precious metal for EOR in acidic media because of its attractive characteristics of high electrochemical activity and low over potential. Unfortunately, high cost and poor anti-poisoning have inhibited Pt-based catalysts application in DEFCs [5,6]. Because of this, many researchers have turned their attention to another metal. Some studies have found that Palladium (Pd) is a more suitable alternative than Pt in DEFC. Although the Pd-based catalyst activity in the acidic medium is not as good as that of the Pt-based catalyst, the Pd-based catalyst performs better than the Pt-based catalyst in the alkaline medium. Regarding price, Pd is cheaper than Pt, which significantly reduces the cost of experimentation and research [7–11].

As we all know, the activity and stability of an electrocatalyst depends significantly on the composition and structure of the supported catalysts and metallic nanoparticles. Therefore, the construction of a new type of Pd-incorporated supported catalyst can play an important role in DEFCs. Carbon-based materials often serve as catalyst supports because of their good electrical conductivity and stability. So far, graphene, carbon microspheres, carbon nanotubes and carbon black have widely been employed as supporting materials of Pd [12–15]. As is well-known, catalyst supports need not only a large specific surface area but also good electrical conductivity, so that the noble metal particles have a high dispersion state while exposing more active sites. Many studies have shown that materials with hollow structures and mesopores can serve as good supports for electrocatalysts. Material such as mesoporous hollow carbon hemispheres [16], spherical carbon capsules [17], hollow graphitized carbon nanocage [18] were used as supports of noble metals electrocatalysts due to their large specific surface area and high stability. Moreover, core-shell structures are widely used in the preparation of catalysts, such as Pd@porous SiO_2 yolk-shell nanostructure [19], $PdO/ZnO@mSiO_2$ [20], $Cu@Cu_2O$ core-shell nanocatalyst [21].

On the other hand, heteroatom-doped carbon materials are considered to be better catalyst supports than pristine carbon-based supports. Studies have indicated that heteroatom doping can effectively improve catalyst activity and make more active sites exposed [22–24]. In particular, when two or more elements, such as N, S, etc. are doped in the structure of carbon, the catalyst activity can improve obviously. That is because N and S as dopants can create synergistic effects that make them easier to modulate for some conjugated bonds and electron distribution, and then making catalysis more efficient [25]. Furthermore, P atom doping could also improve the active sites of the carbon materials [26–28].

In summary, N, S, and P- ternary hollow carbon sphere materials can be used as good precious metal carriers. In this work, N, S, and P- ternary hollow carbon sphere materials (C–N,P,S) were fabricated by a one-step procedure with high-temperature pyrolysis of polyphosphazene coated SiO_2 spheres template composites, followed by removing SiO_2 spheres template via HF solution etching. Finally, Pd nanoparticles were grown on the surfaces of C–N,P,S via the immersion reduction method. The as-prepared Pd/C–N,P,S catalysts exhibited superior electrocatalytic performance toward EOR than that of Pd/C. We have reason to believe that the Pd/C–N,P,S catalyst can be a commercial Pd/C alternative in practical applications towards EOR in the future.

2. Results and Discussion

The morphologies and microstructures of the precursors and Pd/C–N,P,S catalysts were first investigated by scanning electron microscope (SEM) and transmission electron microscope (TEM) images. As shown in Figure 1a, the SEM image of pure SiO_2 microsphere templates exhibits uniform with the diameters ranged from 300 to 500 nm. Through in situ polymerization, the polyphosphazene (PCCP) was coated on the surfaces of SiO_2 microspheres. From the Figure 1b, the obtained PCCP-coated SiO_2 microspheres (PCCP@SiO_2) show the core-shell structures. From the edge, the thickness of the PCCP layer is about 140 nm. Afterward, the PCCP@SiO_2 was pyrolyzed at 900 °C. From the SEM image (Figure 1c), the pyrolyzed PCCP@SiO_2 still remained spherical with a certain fusion. After etching by HF solution, the SiO_2 templates were removed and C–N,P,S were formed. Finally, using C–N,P,S as the supports, the Pd/C–N,P,S was prepared by using in situ reduction. As observed in Figure 1d, Pd/C–N,P,S shows the distinct uniform hollow struture with Pd NPs decorated on the surface. In the EDS elemental mapping (Figure 1e–i), the C, N, P, S, Pd elements are distributed on the surfaces of Pd/C–N,P,S, demonstrating su18ccessful fabrication of Pd/C–N,P,S catalysts.

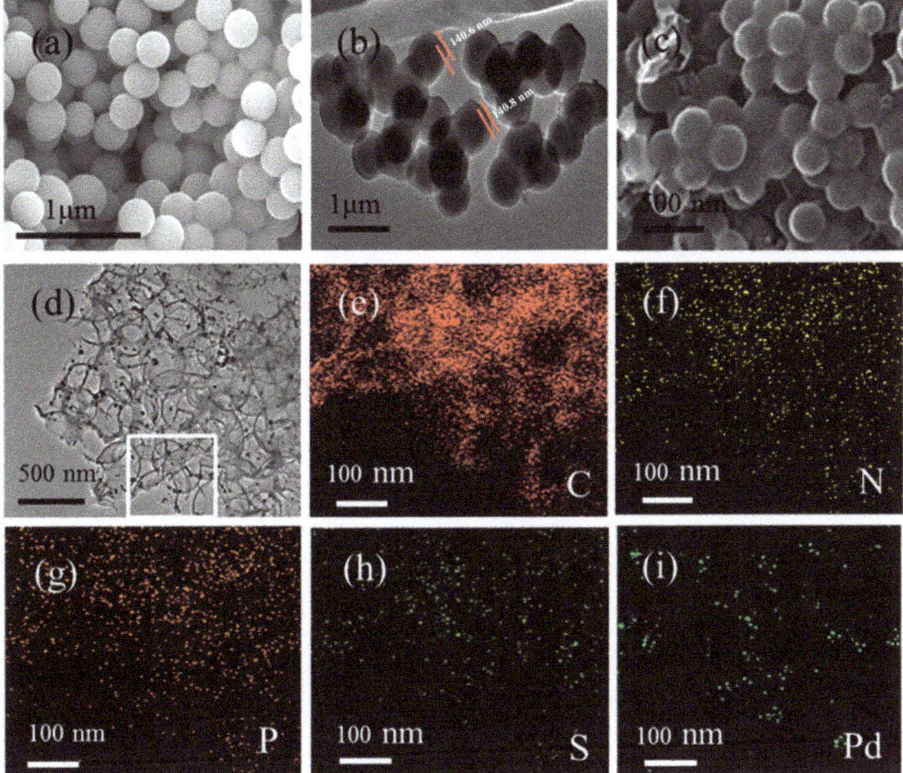

Figure 1. Scanning electron microscope (SEM) images of (**a**) SiO$_2$, transmission electron microscope (TEM) image of (**b**) PCCP@SiO$_2$, (**c**) SEM image of C–N,P,S, (**d**) TEM image of Pd/C–N,P,S and electrospray ionization (EDS) elemental mapping of (**e–i**) C, N, P, S, and Pd elements for Pd/C–N,P,S catalysts.

Furthermore, from the high-resolution TEM (HRTEM) image of the Pd/C–N,P,S (Figure 2a), the metallic Pd NPs with distinct lattice fringe were close contact with the surface of C–N,P,S which benefits the synergistic effect between the Pd and C–N,P,S. The diameter of Pd NPs is ~ 5 nm. The lattice fringe spacing of d = 0.22 nm is attributed to the (111) planes of metallic Pd. Figure 2b shows the X-ray diffraction (XRD) patterns of as-prepared Pd/C and Pd/C–N,P,S. The characteristic peak at 2θ = 25° is corresponded to the C (002) crystal plane. The Pd/C and Pd/C–N,P,S all show characteristic peaks at 2θ values of 39.1°, 45.1°, 66.1°, and 79.7°, which were assigned to (111), (200), (220) and (311) crystal planes of face-centered cubic (fcc) Pd (JCPDS 65–6174), respectively. This agrees with the analysis of the HRTEM image. Figure 3 is the X-ray photoelectron spectroscopy (XPS) spectra of the Pd/C–N, P, S catalyst. The coexistences of C, O, N, P, S, and Pd elements are all in the structure of Pd/C–N,P,S (Figure 3a).The presence of O element may come from the surface hydroxyl groups. In Figure 3b, the high-resolution C 1s XPS spectra were deconvoluted into three distinct peaks which are located at 284.6, 285.5, and 286.7 eV, respectively. The first peak is the classical peak of the sp^{2-} hybridized C–C bond, while the second peak contributes to the C–N or C–S backbone. Besides, the third peak is determined as the C–O groups [23,29]. Furthermore, we also carried out high-resolution XPS analysis of N 1s, P 2p, S 2p and Pd 3d. As shown in Figure 3c, there are two peaks at 398.4 (pyridinic-N) and 402.1 eV (pyridinic-oxide-N) in the deconvoluted N 1s peak [30,31]. The P 2p peak at 133.8 eV can be assigned as the P–N bond (Figure 3d) [26,32]. From the S 2p spectra (Figure 3e), the two peaks at 164.4

and 168.5 eV are contributed to the –S–C–S– and –C–S (O) X –C– groups, which prove that the doped S atoms are tightly bound to the adjacent C atoms [32]. Meanwhile, the high-resolution Pd 3p XPS spectra shown in Figure 3f were deconvoluted into two groups peaks: the peaks at 335.7 and 341 eV are contributed to the metallic Pd. In addition, the other XPS peaks at 337.0 and 342.5 eV could be ascribed to Pd oxide species [33,34].

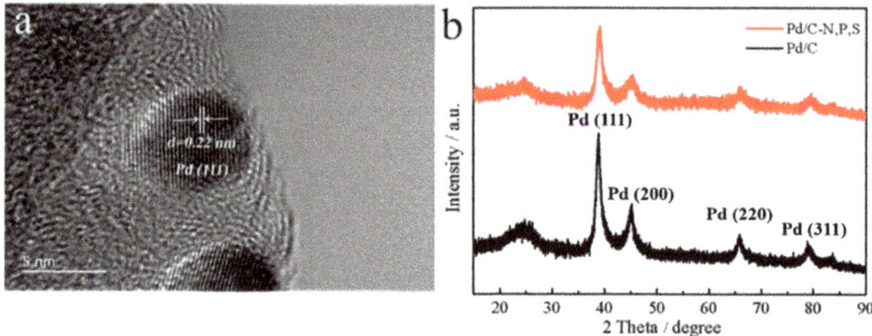

Figure 2. (**a**) The high-resolution TEM (HRTEM) image of the sample Pd/C–N,P,S, (**b**) X-ray diffraction (XRD) patterns of C–N,P,S, Pd/C and Pd/C–N,P,S.

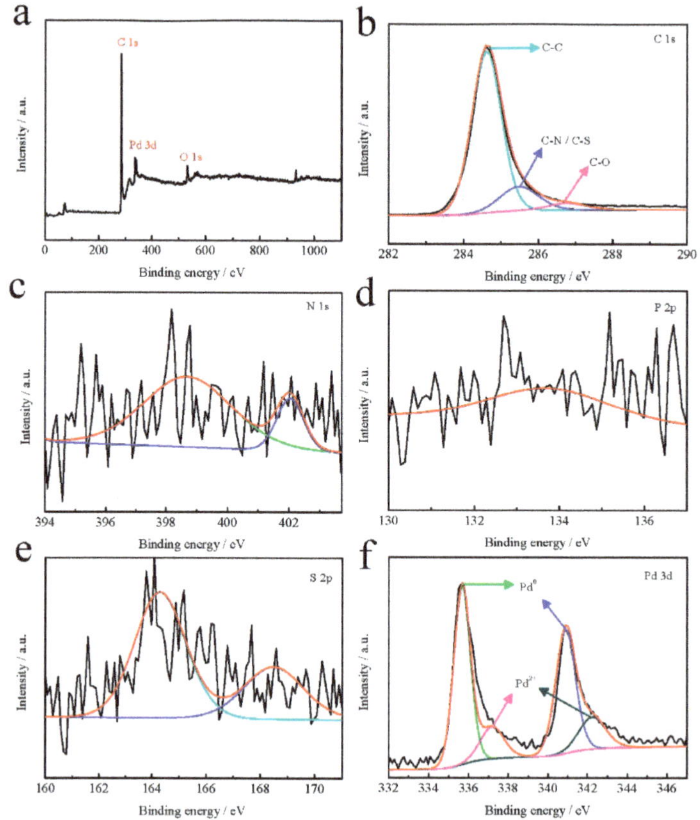

Figure 3. (**a**) X-ray photoelectron spectroscopy (XPS) surveys of the Pd/C–N,P,S catalyst and core-level spectra for (**b**) C 1s, (**c**) N 1s, (**d**) P 2p, (**e**) S 2p and (**f**) Pd 3d.

Figure 4a,b show the N_2 adsorption–desorption isothermals over typical samples of C–N,P,S, and Pd/C–N,P,S. According to the original International Union of Pure and Applied Chemistry (IUPAC) classification, the C–N,P,S, and Pd/C–N,P,S both exhibit type-IV isotherms, and their hysteresis loops are clear H4 type hysteresis loops. The BET specific surface areas of C–N,P,S, and Pd/C–N,P,S are 801 and 212 m^2 g^{-1}, respectively. Furthermore, from the inset BJH pore size and pore size distributions, the pore sizes of the C–N,P,S, and Pd/C–N,P,S are about ~ 4 nm. The large specific surface area of C–N,P,S supports with porous structure can facilitate the infiltration of electrolyte and charge transport in EOR, thereby enhancing the electrochemical activity. As can be seen from Figure 4c, in the Fourier transform–infrared (FT–IR) spectra of SiO_2, PCCP@SiO_2 and pyrolytic PCCP@ SiO_2 composites, the characteristic peaks at 464, 795, and 1092cm^{-1} are attributed to the bending and symmetrical stretching vibrations of Si–O–Si, respectively. Besides, the peak at 952 cm^{-1} is attributed to the specific vibration of Si–O. The peak at 3453 cm^{-1} is assigned to the vibration of hydroxyl groups on the surfaces of SiO_2 microspheres. As templates, the PCCP@SiO_2 with PCCP is coated on the surfaces of SiO_2 microspheres. Toward PCCP@SiO_2, the absorption peaks at 1588 and 1490 cm^{-1} originate from the benzene ring in PCCP. The peaks at 1295 and 1154 cm^{-1} are ascribed to the sulfone in PCCP. Furthermore, the peaks at 1187 and 887 cm^{-1} originate from the vibrations of P=N and P–N in the structure of PCCP. More importantly, the characteristic peak at 949 cm^{-1} implied the formation of P–O–benzene ring. Therefore, the PCCP can be grafted onto the surfaces of SiO_2 microspheres via in situ polymerization. As for sample pyrolytic PCCP@SiO_2, the characteristic peaks of PCCP disappeared. The peaks at 470 and 800 cm^{-1} are corresponded to the symmetrical stretching vibration of Si–O. It demonstrates that the layer of PCCP in the PCCP@ SiO_2 has already changed into porous carbon structure. Figure 4d shows the Raman spectra for the samples of C–N,P,S and Pd/C–N,P,S composites. The D and G bands of carbon were observed at ~1335 and ~1595 cm^{-1} [35]. The value of I_D/I_G for the C–N,P,S ($\approx$1.1) was lower than that of the Pd/C–N,P,S ($\approx$1.18), which could be due to increased defects in C atom for the Pd/C–N,P,S after Pd NPs decoration.

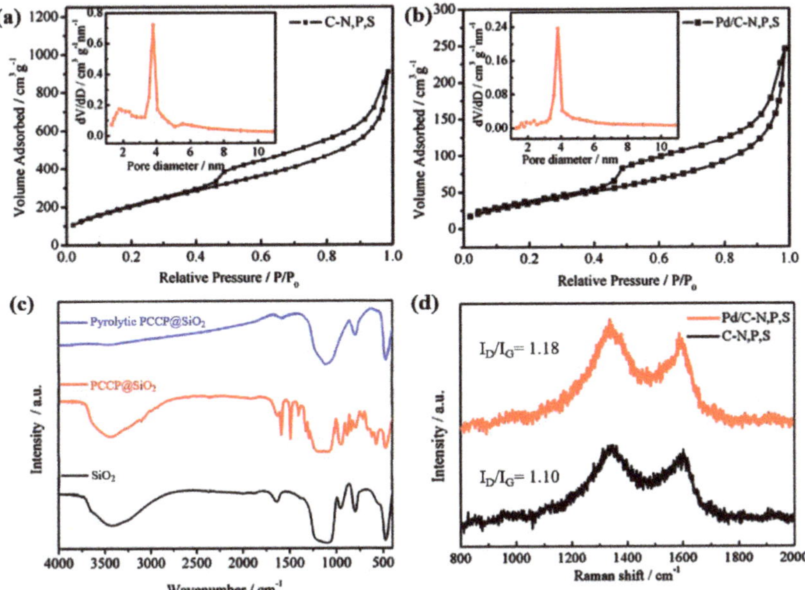

Figure 4. (a) N_2 adsorption–desorption isotherm images of C–N,P,S and (b) Pd/C–N,P,S and pore-size distributions, (c) Fourier transform–infrared (FT–IR) spectra for SiO_2, PCCP@SiO_2 and pyrolytic PCCP@ SiO_2, (d) Raman spectra for the samples of C–N,P,S and Pd/C–N,P,S.

The cyclic voltammetry (CV) curves of Pd/C and Pd/C–N,P,S catalysts were measured in 1 mol/L sodium hydroxide water solution. As shown in Figure 5a, the hydrogen adsorption–desorption peak of Pd/C–N,P,S catalyst shows a higher current density between −1.0 and −0.7 V vs. SCE. Besides, the peaks which is located at around −0.35 V is the reduction peak of PdO_x. It is noting that there is a negative shift of ~ 2 mV for the Pd/C–N,P,S catalyst in comparison with Pd/C, indicating that the decreased oxophilicity on the Pd surfaces can reduce chemisorptions of these oxygen species, which benefits the EOR in the alkaline condition. Furthermore, the CV curves of Pd/C–N,P,S and Pd/C in the sodium hydroxide and ethanol mixture solution were shown in Figure 5b. Obviously, the mass peak current density for Pd/C–N,P,S catalyst can reach ~1686 mA mg^{-1}, which is 2.8 times higher than that of Pd/C. Also, the mass peak current density of Pd/C–N,P,S catalyst is higher than previously reported Pd-based EOR catalysts (Table S1), such as Pd–Ag nanoparticles, Pd_7/Ru_1,etc. [36–42]. Considering the structure of the Pd/C–N,P,S catalyst, the doped N, P, and S atoms can significantly tune the electronic structure of the carbon shell of C–N,P,S. On the one hand, the C–N,P,S with large surface area are conductive to the diffusion of electrolyte and charge transportation; on the other hand, the electrons can transfer from Pd to C–N,P,S due to the existences of heteroatoms. The synergic effects between Pd and C–N,P,S contributed the high electrochemical activity. More importantly, the C–N,P,S supports also can faciliate the CO_{ad} intermediates.

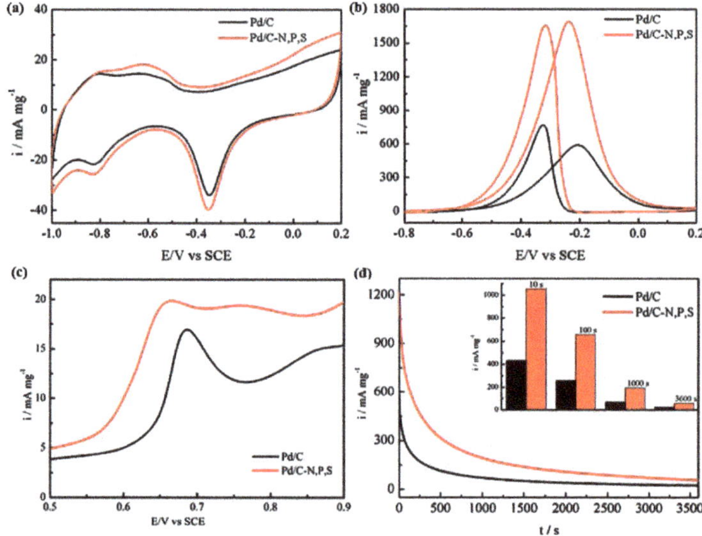

Figure 5. (**a**) Cyclic voltammetry (CV) of Pd/C and Pd/C–N,P,S catalysts in sodium hydroxide (1 mol/L), (**b**) CVs of Pd/C and Pd/C–N,P,S catalysts in sodium hydroxide (1 mol/L) and ethanol (1 mol/L) mixture solution, (**c**) CO monolayer stripping voltammograms of the Pd/C and Pd/C–N,P,S catalysts in diluted sulfuric acid solution (0.5 mol/L), (**d**) chronoamperometric curves for the the Pd/C and Pd/C–N,P,S catalysts in sodium hydroxide (1 mol/L) and ethanol (1 mol/L) mixture solution at −0.20 V. Scan rate: 50 mV/s.

The CO monolayer stripping curves of the Pd/C and Pd/C–N,P,S catalysts were recorded in 0.5 mol/L of sulfuric acid. As shown in Figure 5c, the onset potential of CO oxidation on Pd/C–N,P,S catalyst is 0.665 V. In comparison with Pd/C, there is a negative shift of 2.1 mV, indicating the CO_{ads} can be oxidized at a lower potential on Pd/C–N,P,S catalyst. From CO-stripping CV, the electrochemical surface areas (ECSA) can be calculated by CO desorption [43]. Considering the 420 μC cm^{-2} is the charge required to oxidize a monolayer of CO_{ads} on Pd surface [44]. Therefore, the ECSA of the Pd/C–N,P,S catalyst is 76.31 m^2 g^{-1}, which was 2.76 times higher than that of the Pd/C. This indicated that the synergistic effects between the ternary doped hollow carbon sphere and those uniformly dispersed Pd particles which ensure the Pd/C–N,P,S catalyst with large ECSA and better anti-CO poisoning ability. Generally, the stability of an electrocatalyst is a very important factor for further practical application. The chronoamperometric measurement was carried out in the sodium hydroxide and ethanol mixture solution at a constant potential of -0.2 V. As shown in Figure 5c, the Pd/C–N,P,S catalyst always exhibits the higher oxidation current density than that of the Pd/C after different periods. After a repetitive 200 potential cycling tests, the peak current density of Pd/C–N, P, S has decreased by ~4.2%, and the peak current density of Pd/C has decreased by ~18.6%. Obviously, the Pd/C–N, P, S shows higher structural stability than that of Pd/C. From Figure 5d, towards the ratio (i_{3600}/i_{10}) of the peak current density after 3600 s to that after 10 s, the i_{3600}/i_{10} of Pd/C–N, P, S is 8.6% higher than that of Pd/C (6.7%).

3. Materials and Methods

3.1. Materials

Tetraethoxysilane (TEOS), hexachlorocyclotriphos (HCCP), 4,4-sulfonyldiphenol (BPS), triethylamine (TEA), ethylenediaminetetraacetic acid (EDTA), acetonitrile, sodium carbonate, potassium tetrachloropalladate (II) (99.95%), sodium tetrachloropalladate (99.95%) were all obtained from Shanghai Aladdin Bio-Chem Technology Co.,Ltd (Shanghai, China). Ammonia water and ethanol were purchased from the Sinopharm Chemical Reagent Co. Ltd. (Shanghai, China). All chemicals and solvents were of analytical grade and used without further purification.

3.2. Synthesis of C–N,P,S

Figure 6 shows the schematic illustration of fabrication for C–N,P,S hollow microsphere. Uniform SiO_2 microsphere templates were fabricated by a reported method [45]. First, 0.2 g of SiO_2 was dispersed into 200 mL of acetonitrile. Then, a certain amount of HCCP, BPS and TEA were added to the above SiO_2/acetonitrile dispersion with continuous stirring. After 6 h of stirring, the products of PCCP@SiO_2 was separated by centrifuge washing with water and ethanol, and then vacuum dried at 50 °C for 12 h. Then, PCCP@SiO_2 was subjected to pyrolysis under a nitrogen atmosphere. The pyrolysis condition is 600 °C for 1 h, and then 900 °C for 3 h (2 °C/min). After being pyrolyzed, the C–N, P, S@SiO_2 was collected. IThen, the above C–N, P, S@SiO_2 was etched in HF to remove SiO_2 microsphere templates. Finally, the C–N, P, S was then washed by lots of water and vacuum dried overnight at 50 °C.

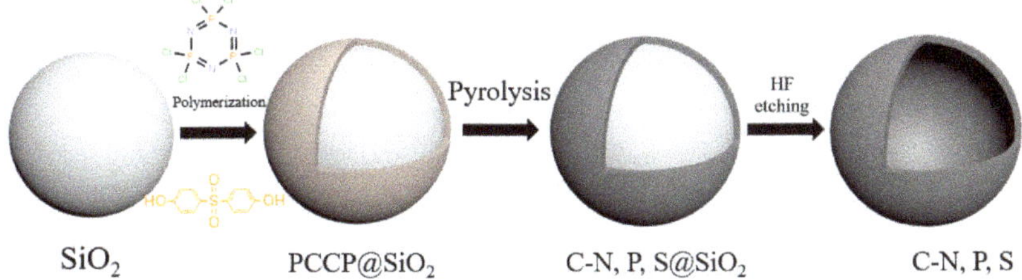

Figure 6. Schematic illustration for fabrication of hollow C–N,P,S sphere.

3.3. Preparation of Pd/C–N,P,S Catalysts

The Pd/C–N,P,S was prepared by a simple and efficient chemical reduction impregnation method. 50 mg of C–N,P,S was first dispersed into 30 mL of H_2O with bath sonication of 30 min. Then, 50 mg of K_2PdCl_4 and 50 mg of EDTA were added into a C–N,P,S/water dispersion with continuous stirring. In the next stage, the pH value of above mixture solution was adjusted to about 10 by concentrated ammonia water. After that, 0.1 g of Na_2CO_3 and 0.1 g of $NaBH_4$ were dissolved in 20 mL of H_2O as reductant solution. The reductant solution was slowly added dropwise to the previous dispersion. After reaction of 4 h with stirring, the Pd/C–N,P,S was collected by centrifugation. After washing by water and ethanol, the Pd/C–N,P,S was vacuum dried at 70 °C for 12 h.

3.4. Electrocatalytic Activity Test

A CHI 660E electrochemistry workstation (Shanghai, China) was used for testing the electrocatalytic performance of all catalysts with a conventional three-electrode method. The graphite rod electrode was employed as counter electrode. A saturated calomel electrode (SCE) was the reference electrode. The Pd/C–N,P,S and Pd/C modified glassy carbon electrodes (Φ = 3 mm) were used as the working electrodes. CO stripping experiment was peformed in diluted sulfuric acid solution. CV tests and chronoamperometry were undertaken in the sodium hydroxide and ethanol mixture solution.

3.5. Catalysts Characterization

X-ray diffraction (XRD) patterns were obtained from a Bruker D8 Advance X-ray diffractometer (Karlsruhe, Germany). The morphology and microstructure of all catalysts were characterized by field-emission scanning electron microscopy (FE-SEM, JSM-7800F, Tokyo, Japan) and transmission electron microscopy (TEM, JEOL JEM-2100F electron microscope, Tokyo, Japan). XPS measurements were performed on a Kratos Axis Ultra DLD (Kratos Analytical, Manchester, UK) with an Al K_α X-ray (1486.6 eV). Fourier transform–infrared spectroscopy (FT–IR) analyses were obtained with a FTIR-8400S spectrometer (Kratos Analytical, Manchester, UK).

4. Conclusions

In summary, the Pd/C–N,P,S with Pd NPs decorated on the surfaces of ternary N, S, and P-doped hollow carbon microspheres was successfully prepared using the SiO_2 microspheres and PCCP as templates and carbon source. As EOR catalyst, the Pd/C–N,P,S exhibits high electrochemical activity with a mass peak current density of 1686 mA mg^{-1} which is 2.8 times higher than that of Pd/C. Remarkably, the ECSA of the Pd/C–N,P,S catalyst can reach ~76.31 m^2 g^{-1} which is 2.76 times higher that of the Pd/C. In addition, the Pd/C–N,P,S catalyst also shows good stability. The high performance of Pd/C–N,P,S catalysts may be mainly due to the following reasons: (1) the porous C–N,P,S with large surface area are conductive to the diffusion of electrolyte and charge transportation, and (2) the electrons can transfer from Pd to C–N,P,S due to the existences of heteroatoms. Also, the C–N,P,S supports can also faciliate the CO_{ad} intermediates. The synergic effects between Pd and C–N,P,S contributed to the high electrochemical activity. Therefore, the Pd/C–N,P,S catalyst has promising prospects for high-performance EOR catalysts.

Supplementary Materials: The following are available online at http://www.mdpi.com/2073-4344/9/2/114/s1: Figure S1: CVs of Pd/C–N,P,S catalysts in sodium hydroxide (1 mol/L) at different scan rates; Figure S2: The calibration plot of oxidation and reduction peaks currents vs. scan rate; Figure S3: CVs of Pd/C-N,P,S and Pd/C after 200 repetitive potential cycling tests, Table S1: The mass peak current densities of various reported EOR catalysts.

Author Contributions: Conceptualization, J.F., Q.L., P.S., Q.X., Y.M.; methodology, J.F., Q.L., P.S., Q.X., Y.M.; software, J.F., Q.L., P.S., Q.X., Y.M.; validation, K.Y., Y.L., J.F., Q.L., P.S., Q.X., Y.M.; formal analysis, K.Y., Y.L.; investigation, K.Y., Y.L., J.F., Q.L., P.S., Q.X., Y.M.; resources, K.Y., Y.L., J.F., Q.L., P.S., Q.X., Y.M.; data curation, K.Y., Y.L., J.F., Q.L., P.S., Q.X., Y.M.; writing—original draft preparation, K.Y., Y.L., Q.X.; writing—review and editing, J.F., Q.L., P.S., Q.X., Y.M.; supervision, Q.L., P.S., Y.M., J.F.; project administration, J.F., Q.X., Y.M.; funding acquisition, J.F., Q.X., Y.M. All authors discussed the results and commented on the manuscript.

Funding: This research was funded by the National Natural Science Foundation of China (Nos.91745112, 21473039, and 21604051). This work was also supported by the Shanghai Municipal Education Commission (Nos. 15ZZ088 and 15SG49), and the Science and Technology Commission of Shanghai Municipality (18020500800).

Conflicts of Interest: The authors declare no conflict of interest.

References

1. Antolini, E.; Gonzalez, E.R. Alkaline direct alcohol fuel cells. *J. Power Sources* **2010**, *195*, 3431–3450. [CrossRef]
2. Antolini, E. Palladium in fuel cell catalysis. *Energy Environ. Sci.* **2009**, *2*, 915–931. [CrossRef]
3. Bianchini, C.; Shen, P.K. Palladium-Based Electrocatalysts for Alcohol Oxidation in Half Cells and in Direct Alcohol Fuel Cells. *Chem. Rev.* **2009**, *109*, 4183–4206. [CrossRef] [PubMed]
4. Martins, C.A.; Fernandez, P.S.; Lima, F.D.; Troiani, H.E.; Martins, M.E.; Arenillas, A.; Maia, G.; Camara, G.A. Remarkable electrochemical stability of one-step synthesized Pd nanoparticles supported on grapheme and multi-walled carbon nanotubes. *Nano Energy* **2014**, *9*, 142–151. [CrossRef]
5. Dutta, A.; Ouyang, J.Y. Ternary NiAuPt nanoparticles on reduced graphene oxide as catalysts toward the electrochemical oxidation reaction of ethanol. *ACS Catal.* **2015**, *5*, 1371–1380. [CrossRef]
6. Zadick, A.; Dubau, L.; Sergent, N.; Berthome, G.; Chatenet, M. Huge instability of Pt/C catalysts in alkaline medium. *ACS Catal.* **2015**, *5*, 4819–4824. [CrossRef]
7. Sneed, B.T.; Young, A.P.; Jalalpoor, D.; Golden, M.C.; Mao, S.; Jiang, Y.; Wang, Y.; Tsung, C.K. Shaped Pd-Ni-Pt Core-Sandwich-Shell Nanoparticles: Influence of Ni Sandwich Layers on Catalytic Electrooxidation. *ACS Nano* **2014**, *7*, 7239–7250. [CrossRef]
8. Wang, E.D.; Xu, J.B.; Zhao, T.S. Density Functional Theory Studies of the Structure Sensitivity of Ethanol Oxidation on Palladium Surfaces. *J. Phys. Chem. C* **2010**, *114*, 10489–10497. [CrossRef]
9. Wang, Y.; Shi, F.; Yang, Y.; Cai, W. Carbon supported Pd-Ni-P nanoalloy as an efficient catalyst for ethanol electro-oxidation in alkaline media. *J. Power Sources* **2013**, *243*, 369–373. [CrossRef]
10. Ahmed, M.S.; Jeon, S.W. Highly Active Graphene-Supported NixPd100–x Binary Alloyed Catalysts for Electro-Oxidation of Ethanol in an Alkaline Media. *ACS Catal.* **2014**, *4*, 1830–1837. [CrossRef]
11. Wang, L.; Lavacchi, A.; Bevilacqua, M.; Bellini, M.; Fornasiero, P.; Filippi, J.; Innocenti, M.; Marchionni, A.; Miller, H.A.; Vizza, F. Energy efficiency of alkaline direct ethanol fuel cells employing nanostructured palladium electrocatalysts. *ChemCatChem* **2015**, *7*, 2214–2221. [CrossRef]
12. Singh, R.N.; Awasthi, R. Graphene support for enhanced electrocatalytic activity of Pd for alcohol oxidation. *Catal. Sci. Technol.* **2011**, *1*, 778–783. [CrossRef]
13. Lin, Y.; Liu, Q.; Fan, J.C.; Liao, K.X.; Xie, J.W.; Liu, P.; Chen, Y.H.; Min, Y.L.; Xu, Q.J. Highly dispersed palladium nanoparticles on poly (N_1, N_3-dimethylbenzimidazolium) iodide-functionalized multiwalled carbon nanotubes for ethanol oxidation in alkaline solution. *RSC Adv.* **2016**, *6*, 102582–102594. [CrossRef]
14. Chang, J.F.; Feng, L.G.; Liu, C.P.; Xing, W.; Hu, X.L. An effective Pd–Ni_2P/C anode catalyst for direct formic acid fuel cells. *Angew. Chem. Int. Ed.* **2014**, *53*, 122–126. [CrossRef] [PubMed]
15. Chen, A.; Ostrom, C. Palladium-based nanomaterials: Synthesis and electrochemical applications. *Chem. Rev.* **2015**, *115*, 11999–12044. [CrossRef] [PubMed]
16. Yan, Z.; Meng, H.; Shi, L. Synthesis of mesoporous hollow carbon hemispheres as highly efficient Pd electrocatalyst support for ethanol oxidation. *Electrochem. Commun.* **2010**, *12*, 689–692. [CrossRef]
17. Iijima, S. Helical microtubules of graphitic carbon. *Nature* **1991**, *354*, 56–58. [CrossRef]
18. Zhang, Q.; Jiang, L.; Wang, H. Hollow graphitized carbon nanocage supported Pd catalyst with excellent electrocatalytic activity for ethanol oxidation. *ACS Sustain. Chem. Eng.* **2018**, *6*, 7507–7514. [CrossRef]
19. Kim, A.; Bae, H.S.; Park, J.C.; Song, H. Surfactant-free pd@pSiO_2 yolk-shell nanocatalyst for selective oxidation of primary alcohols to aldehydes. *New J. Chem.* **2015**, *39*, 8153–8157. [CrossRef]
20. Jinwoo, K.; Aram, K.; Nallal, M.; Kang, P. PdO/ZnO@mSiO_2 hybrid nanocatalyst for reduction of nitroarenes. *Catalysts* **2018**, *8*, 280.
21. Aram, K.; Nallal, M.; Chohye, Y.; Sang, J.; Kang, P. MOF-derived Cu@Cu_2O nanocatalyst for oxygen reduction reaction and cycloaddition reaction. *Nanomaterials* **2018**, *8*, 138.
22. Choi, C.H.; Park, S.H.; Woo, S.I. Binary and ternary doping of nitrogen, boron, and phosphorus into carbon for enhancing electrochemical oxygen reduction activity. *ACS Nano* **2012**, *8*, 7084–7091. [CrossRef] [PubMed]
23. Jiang, Z.; Jiang, Z.J.; Maiyalagan, T.; Manthiram, A. Cobalt oxide-coated N- and B-doped graphene hollow spheres as a bifunctional electrocatalyst for oxygen reduction and oxygen evolution reactions. *J. Mater. Chem. A* **2016**, *4*, 5877–5889. [CrossRef]

24. Liu, M.; Song, Y.; He, S.; Tjiu, W.W.; Pan, J.; Xia, Y.; Liu, T. Nitrogen-doped graphene nanoribbons as efficient metal-free electrocatalysts for oxygen reduction. *ACS Appl. Mater. Interfaces* **2014**, *6*, 4214–4222. [CrossRef] [PubMed]

25. Cazetta, A.L.; Zhang, T.; Silva, T.L.; Almeida, V.C.; Asefa, T. Bone char-derived metal-free N- and S-co-doped nanoporous carbon and its efficient electrocatalytic activity for hydrazine oxidation. *Appl. Catal. B* **2018**, *225*, 30–39. [CrossRef]

26. Choi, C.H.; Chung, M.W.; Park, S.H.; Woo, S.I. Additional doping of phosphorus and or sulfur into nitrogen-doped carbon for efficient oxygen reduction reaction in acidic media. *Phys. Chem. Chem. Phys.* **2013**, *15*, 1802–1805. [CrossRef] [PubMed]

27. Liu, Q.; Fan, J.C.; Min, Y.L.; Wu, T.; Lin, Y.; Xu, Q.J. B, N-codoped graphene nanoribbons supported Pd nanoparticles for ethanol electrooxidation enhancement. *J. Mater. Chem. A* **2016**, *4*, 4929–4933. [CrossRef]

28. Sanchez, M.L.; Primo, A.; Garcia, H. P-doped graphene obtained by pyrolysis of modified alginate as a photocatalyst for hydrogen generation from water–methanol mixtures. *Angew. Chem.* **2013**, *52*, 11813–11816. [CrossRef] [PubMed]

29. Choi, C.H.; Chung, M.W.; Kwon, H.C.; Park, S.H.; Woo, S.I. N- and P, N-doped graphene as highly active catalysts for oxygen reduction reactions in acidic media. *J. Mater. Chem. A* **2013**, *1*, 3694–3699. [CrossRef]

30. Wei, W.; Wang, Q.; Zhang, L.; Liang, H.W.; Chen, P.; Yu, S.H. N-, P- and Fe-tridoped nanoporous carbon derived from plant biomass: An excellent oxygen reduction electrocatalyst for zinc-air battery. *J. Mater. Chem. A* **2016**, *4*, 8602–8609.

31. Zhang, G.G.; Zang, S.H.; Wang, X.C. Layered Co(OH)$_2$ deposited polymeric carbon nitrides for photocatalytic water oxidation. *ACS Catal.* **2015**, *5*, 941–947. [CrossRef]

32. Zhang, C.Z.; Nasir, M.; Yin, H.; Liu, F.; Hou, Y.L. Synthesis of phosphorus-doped graphene and its multifunctional applications for oxygen reduction reaction and lithium ion batteries. *Adv. Mater.* **2013**, *25*, 4932–4937. [CrossRef] [PubMed]

33. Yang, G.; Chen, Y.; Zhou, Y.; Tang, Y.; Lu, T. Preparation of carbon supported Pd–P catalyst with high content of element phosphorus and its electrocatalytic performance for formic acid oxidation. *Electrochem. Commun.* **2010**, *12*, 492–495. [CrossRef]

34. Zhang, X.; Zhu, J.X.; Chandra, S.T.; Ma, Z.Y.; Huang, H.J.; Zhang, J.F.; Lu, Z.Y.; Huang, W.; Wu, Y.P. Palladium nanoparticles supported on nitrogen and sulfur dual-doped graphene as highly active electrocatalysts for formic acid and methanol oxidation. *ACS Appl. Mater. Interfaces* **2016**, *8*, 10858–10865. [CrossRef] [PubMed]

35. Fu, J.W.; Xu, Q.; Chen, J.F.; Chen, Z.M.; Huang, X.B.; Tang, X.Z. Controlled fabrication of uniform hollow core porous shell carbon spheres by the pyrolysis of core/shell polystyrene/cross-linked polyphosphazene composites. *Chem. Commun.* **2010**, *46*, 6563–6565. [CrossRef] [PubMed]

36. Hong, J.W.; Kim, Y.; Wi, D.H.; Lee, S.; Lee, S.-U.; Lee, Y.W.; Choi, S.-I.; Han, S.W. Ultrathin Free-Standing Ternary-Alloy Nanosheets. *Angew. Chem. Int. Ed.* **2016**, *55*, 2753–2758. [CrossRef] [PubMed]

37. Shi, Q.; Zhang, P.; Li, Y.; Xia, H.; Wang, D.; Tao, X. Synthesis of open-mouthed, yolk–shell Au@AgPd nanoparticles with access to interior surfaces for enhanced electrocatalysis. *Chem. Sci.* **2015**, *6*, 4350–4357. [CrossRef]

38. Ye, S.-H.; Feng, J.-X.; Li, G.-R. Pd Nanoparticle/CoP Nanosheet Hybrids: Highly Electroactive and Durable Catalysts for Ethanol Electrooxidation. *ACS Catal.* **2016**, *6*, 7962–7969. [CrossRef]

39. Zhang, K.; Bin, D.; Yang, B.; Wang, C.; Ren, F.; Du, Y. Ru-assisted synthesis of Pd/Ru nanodendrites with high activity for ethanol electrooxidation. *Nanoscale* **2015**, *7*, 12445–12451. [CrossRef]

40. Wang, A.-L.; He, X.-J.; Lu, X.-F.; Xu, H.; Tong, Y.-X.; Li, G.-R. Palladium–Cobalt Nanotube Arrays Supported on Carbon Fiber Cloth as High-Performance Flexible Electrocatalysts for Ethanol Oxidation. *Angew. Chem. Int. Ed.* **2015**, *54*, 3669–3673. [CrossRef]

41. Chen, H.; Huang, Y.; Tang, D.; Zhang, T.; Wang, Y. Ethanol oxidation on Pd/C promoted with CaSiO$_3$ in alkaline medium. *Electrochim. Acta* **2015**, *158*, 18–23. [CrossRef]

42. Li, L.; Chen, M.; Huang, G.; Yang, N.; Zhang, L.; Wang, H.; Liu, Y.; Wang, W.; Gao, J. A green method to prepare Pd–Ag nanoparticles supported on reduced graphene oxide and their electrochemical catalysis of methanol and ethanol oxidation. *J. Power Sources* **2014**, *263*, 13–21. [CrossRef]

43. Spendelow, J.S.; Lu, G.Q.; Kenis, P.J.A.; Wieckowski, A. Electrooxidation of adsorbed CO on Pt(111) and Pt(111)/Ru in alkaline media and comparison with results from acidic media. *J. Electroanal. Chem.* **2004**, *568*, 215–224. [CrossRef]

44. Zhao, Y.; Yang, X.; Tian, J.; Wang, F.; Zhan, L. A facile and novel approach toward synthetic polypyrrole oligomers functionalization of multi-walled carbon nanotubes as PtRu catalyst support for methanol electro-oxidation. *J. Power Sources* **2010**, *195*, 4634–4640. [CrossRef]
45. Stöber, W.; Fink, A.; Bohn, E. Controlled growth of monodisperse silica spheres in the micron size range. *J. Colloid Interface Sci.* **1968**, *26*, 62–69. [CrossRef]

Article

Metallosupramolecular Polymer Precursor Design for Multi-Element Co-Doped Carbon Shells with Improved Oxygen Reduction Reaction Catalytic Activity

Yuzhe Wu, Yuntong Li, Jie Mao, Haiyang Wu, Tong Wu, Yaying Li, Birong Zeng, Yiting Xu, Conghui Yuan * and Lizong Dai *

Fujian Provincial Key Laboratory of Fire Retardant Materials, College of Materials, Xiamen University, Xiamen 361005, China; wuyuzhe@stu.xmu.edu.cn (Y.W.); lyt@stu.xmu.edu.cn (Y.L.); mao-jie@foxmail.com (J.M.); why@stu.xmu.edu.cn (H.W.); wutong@stu.xmu.edu.cn (T.W.); liyaying@stu.xmu.edu.cn (Y.L.); brzeng@xmu.edu.cn (B.Z.); xyting@xmu.edu.cn (Y.X.)
* Correspondence: yuanch@xmu.edu.cn (C.Y.); lzdai@xmu.edu.cn (L.D.); Tel.: +86-592-2186178 (C.Y. & L.D.)

Received: 10 December 2018; Accepted: 15 January 2019; Published: 18 January 2019

Abstract: Heteroatom-doped carbon materials have been extensively studied in the field of electrochemical catalysis to solve the challenges of energy shortage. In particular, there is vigorous research activity in the design of multi-element co-doped carbon materials for the improvement of electrochemical performance. Herein, we developed a supramolecular approach to construct metallosupramolecular polymer hollow spheres, which could be used as precursors for the generation of carbon shells co-doped with B, N, F and Fe elements. The metallosupramolecular polymer hollow spheres were fabricated through a simple route based on the Kirkendall effect. The in situ reaction between the boronate polymer spheres and Fe^{3+} could easily control the component and shell thickness of the precursors. The as-prepared multi-element co-doped carbon shells showed excellent catalytic activity in an oxygen reduction reaction, with onset potential (E_{onset}) 0.91 V and half-wave ($E_{half-wave}$) 0.82 V vs reversible hydrogen electrode (RHE). The fluorine element in the carbon matrix was important for the improvement of oxygen reduction reaction (ORR) activity performance through designing the control experiment. This supramolecular approach may afford a new route to explore good activity and a low-cost catalyst for ORR.

Keywords: carbon shell; metallosupramolecular polymer; hollow particles; doping; oxygen reduction reaction

1. Introduction

During the commercialization process of hydrogen fuel cells, exploring electrocatalysts with high oxygen reduction reaction (ORR) activity and outstanding stability is the primary task [1–3]. At present, precious metals incorporating carbon materials, such as commercial Pt/C, have been successfully used as ORR catalysts [4,5]. However, their expensive cost and scarcity greatly limit their broader application [6–8]. Therefore, fabricating non-precious metal composite materials with low-cost, and a catalytic performance comparable or better than commercial Pt/C toward ORR, is of great significance to commercial applications [9,10].

Pure carbon materials have many advantages, including excellent electrical transport properties, a highly active surface area, chemical stability and superior thermal stability. They have thus become an ideal choice for electrochemical energy storage materials [11–15]. However, pure carbon materials have a highly hydrophobic surface and limited active sites, which bring many problems for the application in hydrogen fuel cells. Heteroatom-doping has been recognized as an effective approach to increase the electrochemical activity and surface wettability of carbon materials [16,17]. The chemical elements used for doping carbon materials include nitrogen, phosphorus, boron, fluorine and non-precious metals [18–22]. Notably, N, B co-doped carbon materials have shown excellent ORR performance, because they possess large amounts of defects and active sites [23,24]. Except for the heteroatom-doping materials, carbon materials incorporated with metal and N elements (M–N–C) have been considered as one of the promising candidates. Most of the M–N–C catalysts were prepared via heat-treating the carbon materials with an N-containing compound, as well as metal salts, or obtained from the simple pyrolysis of transition metal macrocyclic polymers [25–27]. Nevertheless, these approaches have shortcomings, such as inhomogeneity of the carbon materials, the aggregation of the active sites, complicated synthesis, as well as high-cost [28,29]. Hence, there exists a need to develop a facile and effective route to fabricate polymeric precursors, containing multiple elements like B, P, S, F, Co and Fe. This is of great importance to incorporate multi-elements into the carbon materials.

We have developed a supramolecular approach, in which the condensation reaction between boronic and catechol monomers is accompanied with the formation of B–N dative bonds, and can organize as formed boronate polymers into nanospheres with controllable sizes [30]. The catechol moiety has a high coordination efficiency with transition metal ions and the Kirkendall effect occurs during the reaction between boronate polymer nanospheres and transition metal ions, thus resulting in the formation of metallosupramolecular polymer hollow spheres [31]. In this work, we extended this approach through the design of the building blocks of the boronate polymers, and therefore metallosupramolecular polymer hollow sphere precursors, containing B, N, F and Fe elements, could be fabricated. Carbonation of the metallosupramolecular polymer hollow sphere precursor at 650 °C afforded carbon shells co-doped with B, N, F and Fe elements. We focused on the control over the thickness of the shell, the influence of the shell thickness and doping elements (B, N, F and Fe) on the ORR performance of the carbon materials.

2. Results and Discussion

2.1. Morphology Evolution

The synthetic process of carbon spheres (CSs) is shown in Scheme 1. We firstly prepared the boronate polymer nanospheres through a simple condensation reaction between 4,4′-((1E,1′E)-(((((perfluoropropane-2,2-diyl)bis(4,1-phenylene))bis(oxy))bis(4,1-phenylene))bis(azanylylidene))bis(methanylylidene))bis(benzene-1,2-diol) (DFC) and (((1E,1′E,1″E)-((nitrilotris(benzene-4,1-diyl)tris(azanylylidene))tris(methanylylidene)) tris(benzene-4,1-diyl))triboronic acid (TBB). The catechol moiety has a high coordination ability to transition metal ions. This coordination interaction is of particular interest in the fabrication of supramolecular polymers, assemblies, smart hydrogels and metal–organic frameworks [32–34]. Since the boronate moiety is dynamic, there probably existed free catechol groups in our boronate polymer nanospheres. Thus, we intended to introduce Fe^{3+} into the boronate polymer nanoparticles through the catechol–Fe^{3+} coordination, thereby forming metallosupramolecular polymers, which comprise Fe, N, B and F elements. After carbonization, carbon materials co-doped with Fe, N, B and F could be obtained.

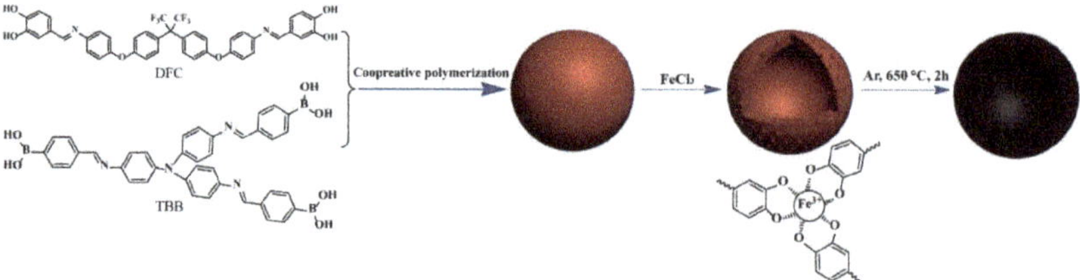

Scheme 1. Synthetic process of the carbon spheres (CSs).

With the assistance of the B–N coordination, a cross-linking reaction between DFC and TBB adopted a typical cooperative polymerization mechanism, and therefore, mono-dispersed boronate polymer nanospheres (BPN) could be formed in one-step (Figure 1a,b) [30]. The coordination reaction between catechol and Fe^{3+} could evidently change the morphology of the boronate polymer nanospheres, which accorded well with a previous report [31]. The metallosupramolecular polymer could be easily carbonized to afford CSs with hollow structures.

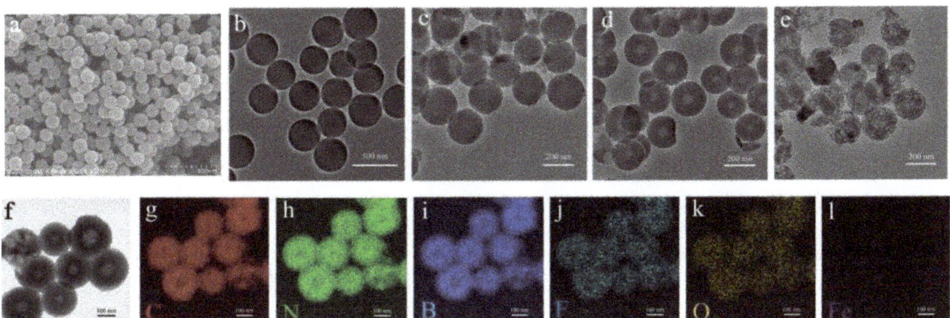

Figure 1. SEM (**a**) and TEM (**b**) images of boronate polymer nanosphere (BPN) nanoparticles. TEM images of $CS_{6\text{-}650}$ (**c**), $CS_{12\text{-}650}$ (**d**) and $CS_{18\text{-}650}$ (**e**). High-angle annular dark-field imaging scanning transmission electron microscope (HAADF-STEM) image (**f**) and corresponding energy-dispersive X-ray (EDX) mappings (**g–l**) of $CS_{12\text{-}650}$.

Figure 1c–e show the TEM images of the CSs prepared from precursors derived from different reaction times. Small pores were created in the interior of the nanospheres within a 6 h reaction time (Figure 1c). When the reaction time was 12 h, hollow nanospheres with shell thickness ~70 nm were obtained (Figure 1d). However, too long of a reaction time was likely to destroy the nanospheres, because many of the obtained particles collapsed (Figure 1e). As demonstrated by the previous report, hollow structural metallosupramolecular polymer precursors were generated according to the Kirkendall effect during the reaction between boronate polymer nanospheres and Fe^{3+} [31]. This process greatly relied on the removal of the boronic acid component. Too long of a reaction time between the boronate polymer nanospheres and Fe^{3+} can result in excessive removal of the boronic acid component, therefore leading to the collapse of the particles. Dark-field TEM imaging of carbon particles obtained from a 12 h reaction time between the boronate polymer nanospheres and Fe^{3+} confirmed the hollow structure (Figure 1f). Energy-dispersive X-ray (EDX) mapping of representative hollow particles indicated the coexistence of C, N, B, F, O and Fe elements (Figure 1g–l). The carbonization temperature likely had no evident effect on the morphology of the carbon materials, as the TEM images of $CS_{12\text{-}550}$ and $CS_{12\text{-}750}$ (Figure S1), display similar morphology to $CS_{12\text{-}650}$.

2.2. Composition and Structure Characterization

Figure 2a gives the Raman spectra of the CSs. The two prominent peaks at 1340 and 1571 cm^{-1} represent the D band of disordered graphitic structure and the G band of ordered carbon matrix, respectively. The calculated intensity ratio from the D band and the G band (I_D/I_G) was applied to evaluate the disorder degree of the carbon materials. The calculated I_D/I_G value of $CS_{6\text{-}650}$, $CS_{12\text{-}650}$ and $CS_{18\text{-}650}$ were 1.22, 1.26 and 1.29, respectively. Probably, the content defect site in the carbon materials increased with the increasing reaction between BPN and Fe^{3+}. The calculated I_D/I_G value of $CS_{12\text{-}550}$ and $CS_{12\text{-}750}$ were 0.94 and 1.19, respectively, which were lower than that of $CS_{12\text{-}650}$.

The crystalline structures of CSs were characterized through XRD. As shown in Figure 2b, an evident broad diffraction peak located at about 25° could be attributed to the (002) plane of ordered graphitic structure. The sharp peak at about 45° was the (110) plane of iron, reduced by the hydrogen–argon mixture gas. Two broad diffraction peaks at about 35° and 43° were derived from the (311) and (222) planes of Fe_3O_4 [35]. With the elongation of reaction time between BPN and Fe^{3+}, the characteristic peaks of both iron and Fe_3O_4 were obviously enhanced. It is likely that penetration of Fe^{3+} into BPN and formation of metallosupramolecular polymers was time dependent. The effect of carbonization temperature on the crystalline structure of CSs was studied by XRD. CSs obtained from different carbonization temperatures generally had similar peak positions. However, with the carbonization temperature increased from 550 °C to 750 °C, the shape, width and intensity of the peak at 25° changed. Therefore, the carbonization temperature could affect the carbon matrix crystalline structure of CSs. It was observed that $CS_{12\text{-}650}$ prepared by the carbonization temperature of 650 °C had the best crystallinity.

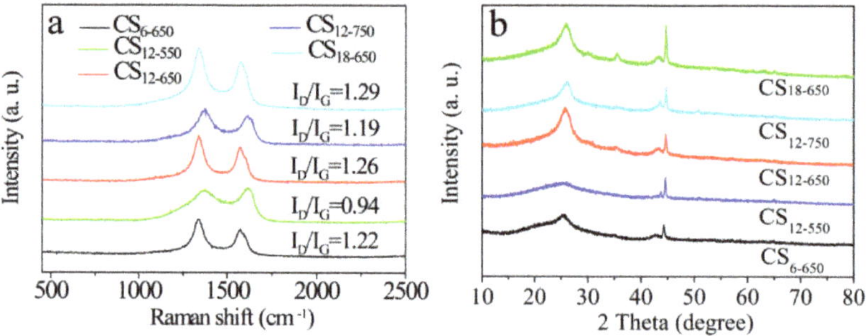

Figure 2. Raman spectra (**a**) and XRD patterns (**b**) of CSs.

The pore character of CSs was tested by physisorption of nitrogen at 77 K. As shown in Figure 3, CSs comprise both microporous and mesoporous structures. Detailed Brunauer–Emmett–Teller (BET) data is also listed in Table 1. The surface areas of $CS_{6\text{-}650}$, $CS_{12\text{-}650}$ and $CS_{18\text{-}650}$ are 361.62, 439.47 and 451.13 m^2 g^{-1}, with relative pore volumes of 0.34, 0.40 and 0.36 cm^3 g^{-1}, respectively. Obviously, their surface areas mainly resulted from the microporous- and mesoporous-pore structures. The specific surface area of the CSs increased gradually with the increase of reaction for the generation of the precursor. The mesoporous-pore volumes of $CS_{6\text{-}650}$, $CS_{12\text{-}650}$ and $CS_{18\text{-}650}$ were 212.08, 258.68 and 190.00 m^2 g^{-1}, respectively. Apparently, $CS_{12\text{-}650}$ had the maximum mesoporous-pore compared with $CS_{6\text{-}650}$ and $CS_{18\text{-}650}$. The mesoporous-pore is important and beneficial for ORR. The surface areas of $CS_{12\text{-}550}$ and $CS_{12\text{-}750}$ were 394.38 and 445.20 m^2 g^{-1}, with relative pore volumes of 0.34 and 0.35 cm^3 g^{-1} (Table 1), respectively, which were lower than that of $CS_{12\text{-}650}$ and $CS_{18\text{-}650}$. This result indicated that the precursor might be not completely carbonized at 550 °C, and a carbonization temperature of 750 °C was not helpful for the development of the pore structure.

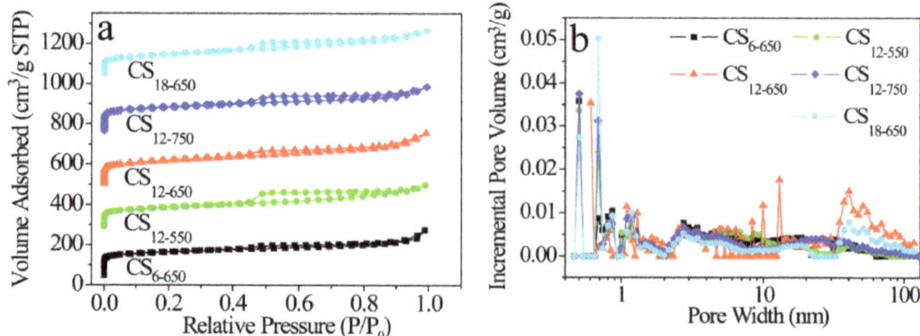

Figure 3. N$_2$ adsorption and desorption isotherms (**a**) and density functional theory (DFT) pore size distribution (**b**) of CSs.

Table 1. Surface area, porosity of CSs.

Samples	S_{BET} [a] [m^2 g^{-1}]	S_{micro} [b] [m^2 g^{-1}]	$S_{meso + macro}$ [c] [m^2 g^{-1}]	V_{total} [d] [cm^3 g^{-1}]
CS$_{6-650}$	361.62	149.54	212.08	0.34
CS$_{12-550}$	394.38	182.54	211.84	0.34
CS$_{12-650}$	439.47	180.79	258.68	0.40
CS$_{12-750}$	445.20	178.97	266.23	0.35
CS$_{18-650}$	451.13	261.13	190.00	0.36

[a] Specific surface area obtained from BET, [b] Surface area of micropores calculated by the t-plot method, [c] Surface area of mesopores and macropores calculated by the t-plot method, [d] Total pore volume.

The X-ray photoelectron spectroscopy (XPS) survey spectra of CS$_{6-650}$, CS$_{12-650}$ and CS$_{18-650}$ are shown in Figure 4a. In addition, the high-resolution XPS spectra of C 1s, N 1s, B 1s, F 1s and Fe 2p of CS$_{12-650}$ were characterized (Figure 4b–f). The C 1s signal can be split into four representative peaks at about 288.2 (C=O), 285.4 (C–O, C–N), 284.4 (C–C, C=C) and 283.6 eV (C–B) (Figure 4b). The six peaks of N 1s spectrum shown in Figure 4c at 403.2, 401.3, 400.3, 399.4, 398.3 and 397.7 eV are attributed to oxidized N, graphitic N, pyrrolic N, amine, pyridinic N, as well as N–B [23]. Notably, pyridinic N was recognized to produce excellent oxygen reduction reaction activity [24]. Three typical peaks of B 1s at 192.1, 190.8 and 189.2 eV were assigned to B–O, B–N, B–C (Figure 4d), separately [36]. The F 1s signal is displayed in Figure 4e. This indicated that fluorine remains in the carbon matrix. The weak peak at 685.7 eV was assigned to the F$^-$ anions. The existence of the F element had the ability to accelerate the oxygen reduction reaction process of carbon catalysts [37]. For the Fe 2p spectrum shown in Figure 4f, the peak at 723.2 eV was assigned to the binding energy of Fe^{2+}, and the peak for Fe^{3+} was detected at 725.4 eV for the $2p_{1/2}$ band. Another two peaks at 714.4 and 710.5 eV could be respectively attributed to the binding energies of the $2p_{3/2}$ orbitals of Fe^{3+} and Fe^{2+} species [38]. The Fe $2p_{3/2}$ peak at approximately 703.5 eV corresponded to Fe^0. The last peak at 719.6 eV was a satellite peak. The XPS results, in combination with the XRD results, clearly confirmed that only a small amount of iron element was transformed into Fe$_3$O$_4$, and most of iron element was changed into zero-valent iron, which may be of potential in improving the electrical conductivity of the carbon materials.

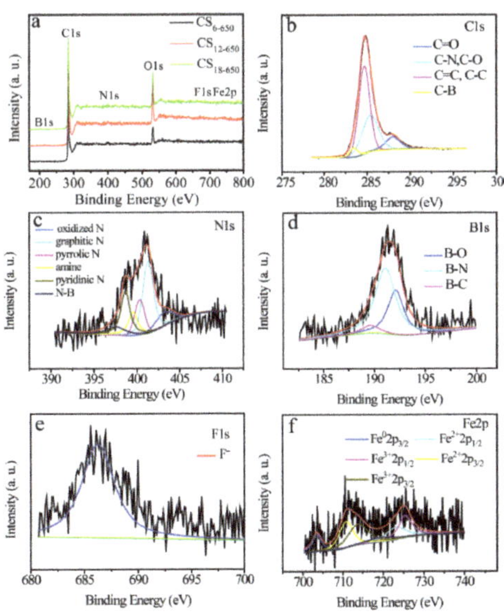

Figure 4. X-ray photoelectron spectroscopy (XPS) survey spectra of CS$_{6-650}$, CS$_{12-650}$ and CS$_{18-650}$ (**a**). High-resolution XPS spectra of (**b**) C 1s, (**c**) N 1s, (**d**) B 1s and (**e**) F 1s, (**f**) Fe 2p for the CS$_{12-650}$.

The XPS survey spectra of $CS_{12\text{-}550}$ and $CS_{12\text{-}750}$ are shown in Figure S2. $CS_{12\text{-}550}$ obtained at 550 °C had F content of 0.94%. However, at carbonization temperature of 750 °C, the prepared $CS_{12\text{-}750}$ comprised no fluorine element (Table S1). Because the C–F bond in the precursor accorded to a regular fracture, the temperature of 650 °C was the highest temperature that could retain the fluorine element in the carbon matrix, as well as endow the materials with adequate carbonization.

2.3. ORR Performances

We then tested the electrochemical performance of the CSs to evaluate their potential as ORR electrocatalysts. The cyclic voltammetry (CV) curves of $CS_{6\text{-}650}$, $CS_{12\text{-}650}$ and $CS_{18\text{-}650}$ were tested in Ar- or O_2-saturated 0.1 M KOH solution (Figure 5a). All samples only displayed an obvious reduction peak in O_2-saturated solution. The linear sweep voltammetry (LSV) curves of CSs are shown in Figure 5b–d. The onset potential (E_{onset}) of $CS_{6\text{-}650}$, $CS_{12\text{-}650}$ and $CS_{18\text{-}650}$ for the oxygen reduction reaction were 0.83 V, 0.91 V and 0.78 V, respectively. The half-wave ($E_{half\text{-}wave}$) potentials of $CS_{6\text{-}650}$, $CS_{12\text{-}650}$ and $CS_{18\text{-}650}$ for the oxygen reduction reaction were 0.79 V, 0.82 V and 0.76 V. This result was also confirmed by a LSV test in the O_2-saturated 0.1 M KOH solution, at a rotation rate of 1600 rpm, with a scan rate of 10 mV/s (Figure 5e). On the other hand, the catalytic activity of the commercial 20 wt% Pt/C was tested under identical experimental conditions (Figure 5e). In comparison, the E_{onset} and $E_{half\text{-}wave}$ of $CS_{12\text{-}650}$ were 0.91 V and 0.82 V, respectively, which were close to that of the commercial 20 wt% Pt/C catalyst with the onset and half-wave potentials of 0.95 V and 0.85 V vs. RHE.

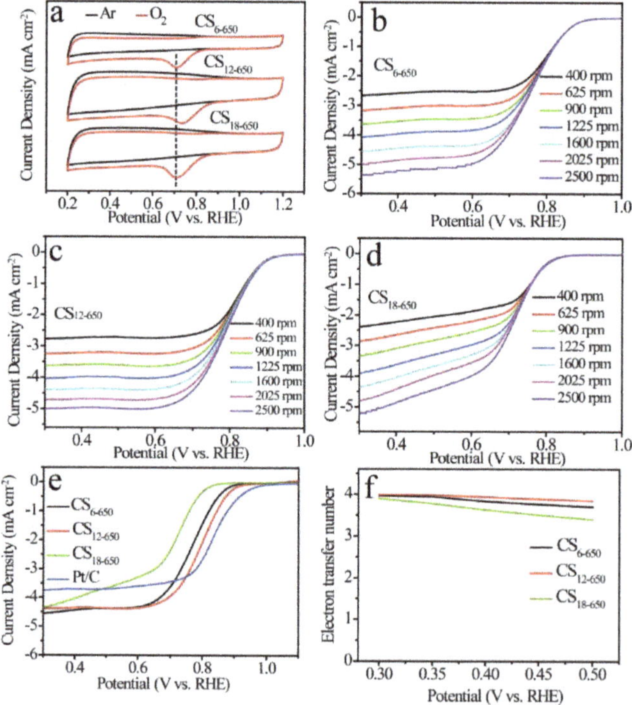

Figure 5. CV curves of $CS_{6\text{-}650}$, $CS_{12\text{-}650}$ and $CS_{18\text{-}650}$ in Ar- or O_2-saturated 0.1 M KOH aqueous solution at a scan rate of 50 mV/s (**a**). Linear sweep voltammetry (LSV) curves of $CS_{6\text{-}650}$ (**b**), $CS_{12\text{-}650}$ (**c**) and $CS_{18\text{-}650}$ (**d**) in O_2-saturated 0.1 M KOH aqueous solution at different rotation speeds. LSV curves of $CS_{6\text{-}650}$, $CS_{12\text{-}650}$ and $CS_{18\text{-}650}$ and the commercial 20 wt% Pt/C in O_2-saturated 0.1 M KOH aqueous solution at a rotation rate of 1600 rpm and a scan rate of 10 mV/s (**e**). Electron transfer numbers obtained from The Kouteckye–Levich (K–L) plots of $CS_{6\text{-}650}$, $CS_{12\text{-}650}$ and $CS_{18\text{-}650}$ (**f**).

The evidently improved electrocatalytic performance of CS_{12-650} could be explained by the following two reasons. First, CS_{12-650} had relatively higher specific surface area compared with CS_{6-650} as proved by BET results, thus leading to the exposure of more active sites. CS_{12-650} also had a higher content of mesoporous-pore than CS_{6-650} and CS_{18-650}, and the mesoporous-pore was beneficial for ORR. Second, CS_{12-650} had relatively thin shell thickness and a regular hollow morphology, which was beneficial for the transfer of water and oxygen during the catalytic reaction. As illustrated by the LSV results, the second platform of the LSV curves of CS_{12-650} became very flat in comparison with CS_{18-650}, indicating the equilibrium of the catalytic reaction. As the morphology of CS_{18-650} was completely damaged and collapsed, the irregular structure could have increased the specific surface area of CS_{18-650} to some extent, but also could have prevented the generation and loading of active sites during carbonization. The activity of CS_{18-650} was thus reduced. Therefore, we consider that a synergistic effect between surface area, shell thickness, morphology and enough available active sites, directly led to the optimization of ORR catalytic activity of CS_{12-650}.

The LSV curves at different rotating rates were tested to study the reaction kinetics of the ORR catalyzed by CSs. The current density of CS_{6-650}, CS_{12-650} and CS_{18-650} increased when increasing the rotating rate from 400 to 2500 rpm, as shown in Figure 5b–d. This could be attributed to the decrease of the diffusion distance. The Kouteckye–Levich (K–L) plots were calculated from the above LSV curves, which revealed a clear linear relationship and quite similar slopes at the corresponding potential, ranging from 0.3 V to 0.5 V (Figure S3a–c). Obviously, the ORR catalyzed by CSs accorded classic first-order reaction kinetics. By calculating underlying scope shown in Figure 5f, the electron transfer numbers (n) derived from the corresponding slopes of Kouteckye–Levich (K–L) plots were approximately 3.94 for CS_{6-650}, 3.96 for CS_{12-650} and 3.88 for CS_{18-650}, verifying an atypical four-electron transfer process. This is the same as the many reported B and N elements co-doped carbon materials [23,24,39].

The CV curves of CS_{12-550} and CS_{12-750} were also tested in Ar- or O_2-saturated 0.1 M KOH solution (Figure 6a), and the corresponding LSV curves of CS_{12-550} and CS_{12-750} are shown in Figure 6b,c. The onset potential (E_{onset}) of CS_{12-550} and CS_{12-750} were 0.85 V and 0.86 V. The half-wave ($E_{half-wave}$) potentials of CS_{12-550} and CS_{12-750} were 0.74 V and 0.77 V. In comparison, the E_{onset} and $E_{half-wave}$ of CS_{12-650} were 0.91 V and 0.82 V, respectively, which were much better than the CS_{12-550} and CS_{12-750}. The electron transfer numbers (n) derived from the corresponding slopes of Kouteckye–Levich (K–L) plots (Figure S3d–f) were approximately 3.26 for CS_{12-550} and 3.56 for CS_{12-750}, verifying a typical four-electron transfer process.

Since they were derived from the same precursor, we considered that the carbonization temperature could evidently affect the ORR performance of the CSs. First, CS_{12-650} had higher I_D/I_G value than CS_{12-550} and CS_{12-750}, implying more defect sites in the carbon matrix. Second, CS_{12-650} had relatively higher specific surface area relative to CS_{12-550}, as proved by BET results, and thus lead to the exposure of more active sites. The specific surface area of CS_{12-650} and CS_{12-750} was nearly the same, so CS_{12-650} and CS_{12-750} had the same utilization towards the active site in the same condition. Third, comparing CS_{12-650} and CS_{12-750}, the existence of the fluorine element in CS_{12-650} was important for the improved ORR performance of CS_{12-650}. Therefore, we considered that the fluorine element in the carbon matrix could promote the generation of defect sites, which was beneficial and important for the ORR activity.

The durability of CS_{12-650} was tested at 1600 rpm in O_2-saturated 0.1 M KOH aqueous solution after 1000 cycles of CV curves (Figure 7a). The onset and half-wave potentials of CS_{12-650} were kept at 0.90 V and 0.80 V. No evident current decrease in the onset and half-wave potentials was observed. The attenuation of electrocatalytic activity was less than 5%, which was much better than the commercial 20 wt% Pt/C catalysts (as shown in Figure S4). Thus, CSs had preferable durability for ORR in the alkaline condition. The relevant crossover effects test was also carried out by taking CS_{12-650} as an example through a chronoamperometric experiment. After adding 3.0 M methanol, the methanol oxidation reaction made the current density of the commercial Pt/C decrease immediately (Figure 7b).

However, the current density of $CS_{12\text{-}650}$ did not display an evident change. These results directly confirmed that $CS_{12\text{-}650}$ has good catalytic selectivity for the oxygen reduction reaction.

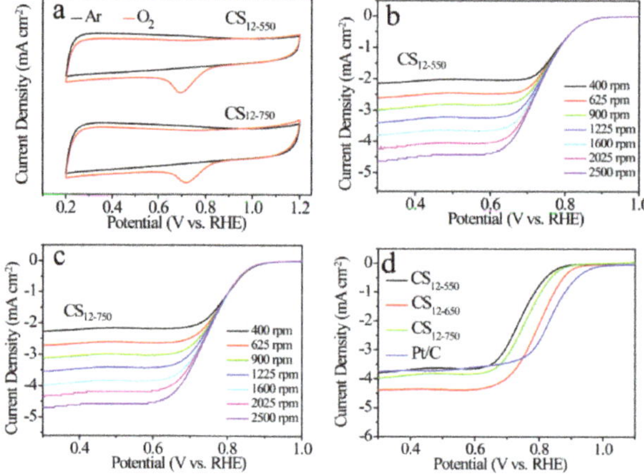

Figure 6. CV curves of $CS_{12\text{-}550}$ and $CS_{12\text{-}750}$ in Ar- or O_2-saturated 0.1 M KOH aqueous solution with a scan rate of 50 mV/s (**a**). LSV curves of $CS_{12\text{-}550}$ (**b**) and $CS_{12\text{-}750}$ (**c**) and (**d**) LSV curves of $CS_{12\text{-}550}$, $CS_{12\text{-}650}$ and $CS_{12\text{-}750}$ and the commercial 20 wt% Pt/C in O_2-saturated 0.1 M KOH aqueous solution at a rotation rate of 1600 rpm and a scan rate of 10 mV/s.

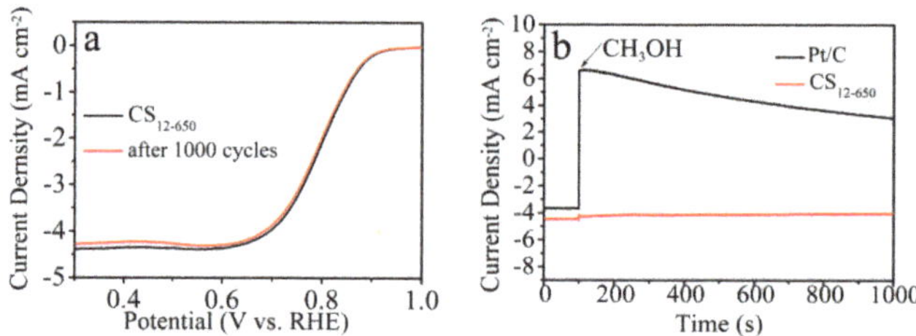

Figure 7. LSV curves of the $CS_{12\text{-}650}$ at 1600 rpm in O_2-saturated 0.1 M KOH aqueous solution before and after 1000 cycles of CV curves with a scan rate of 50 mV/s (**a**). Current-time chronoamperometric responses of the commercial 20 wt% Pt/C and $CS_{12\text{-}650}$ at 0.75 V in O_2-saturated 0.1 M KOH aqueous solution followed by addition of 3.0 M methanol. The rotation rate is 1600 rpm (**b**).

3. Materials and Methods

3.1. Materials

Iron(III) chloride hexahydrate, 4-formylphenylboronic acid and 3,4-dihydroxybenzaldehyde were purchased from Aladdin Company (Shanghai, China) and directly used as received. Tris(4-aminophenyl)amine was obtained from J&K Chemical (Beijing, China). KOH, anhydrous methanol and anhydrous ethanol were supplied by Shanghai Chemical Reagent Industry (Shanghai, China). Nafion (5 wt%) was purchased from Sigma-Aldrich (St. Louis, MO, USA). Boronic monomer TBB was synthesized through our reported method [23]. The catechol monomer DFC was directly synthesized through the Schiff base formation reaction between 4,4′-(((perfluoropropane-2,2-diyl)bis(4,1-phenylene))bis(oxy))dianiline and 3,4-dihydroxybenzaldehyde (detailed procedure is described in Supporting Information).

3.2. Catalyst Preparation

DFC and TBB were dissolved in methanol to afford a concentration of 1.0 mg/mL. TBB solution (6 mL, 0.087 mmol) was added into DFC solution (10 mL, 0.132 mmol) dropwise under N_2 atmosphere with vigorous stirring. The mixture solution became deep orange and a suspension of boronate polymer nanospheres (BPN) formed. BPN powder was obtained by centrifugation and washed with anhydrous methanol three times.

BPN was redispersed in 16 mL of methanol to get 2.0 mg/mL of particle suspension. Then 2.37 mL of methanol solution of $FeCl_3$ (10 mg/mL, 0.037 mmol/mL) was added to the suspension of BPN. After different reaction times, such as 6, 12 and 18 h, the resultant hollow particles were collected by centrifugation, washed by anhydrous methanol, and then dried in vacuum oven at 50 °C overnight. The hollow particles were firstly carbonized at 650 °C for 2 h in an argon atmosphere with a fixed heating rate of 5 °C /min. The second carbonation was performed at 650 °C for 1 h in a mixture gas containing 5% hydrogen and 95% argon with a fixed heating rate of 10 °C /min to reduce Fe^{3+} into Fe, thus forming B, N, F and Fe co-doped hollow carbon nanospheres (denoted as CSs). $CS_{6\text{-}650}$, $CS_{12\text{-}650}$ and $CS_{18\text{-}650}$ represent carbon shells prepared from 6, 12 and 18 h reaction times between BPN and $FeCl_3$, respectively. To evaluate the carbonization temperature on the electrochemical properties of carbon shells, control experiments were performed at pyrolysis temperatures of 550 and 750 °C, and the prepared samples were denoted as $CS_{12\text{-}550}$ and $CS_{12\text{-}750}$.

3.3. Characterization

The scanning electron microscopy (SEM) were characterized using an SU-70 microscope (HITACHI, Tokyo, Japan). The morphology of the samples was tested by transmission electron microscopy (JEM-2100) (JEOL, Tokyo, Japan). The high-angle annular dark-field scanning transmission electron microscopy (HAADF-STEM) images, elemental energy-dispersive X-ray spectroscopy (EDX) mapping and line scanning analyses were acquired on a FEI TECNAI F20 microscope (Hillsboro, OR, USA) tested at 200 kV. The Raman spectra were tested on a Labram HR800 Evolution (Horiba, Lille, France). The powder X-ray diffraction (XRD) patterns were obtained through a desktop X-ray Diffractometer using Cu (600 W) Kα radiation (Rigaku, Tokyo, Japan) to characterize the crystallographic structure of the samples. The Brunauer–Emmett–Teller (BET) surface area and pore volume of the samples were tested through an ASAP 2460 system (Norcross, GA, USA). Before measurement, all of the samples were uniformly degassed at 120 °C for 12 h under vacuum before the test. The X-ray photoelectron spectroscopy (XPS) was tested by PHI Quantum-2000 photoelectron spectrometer (Physical Electronics, Inc., Chanhassen, MN, USA) through a monochromatic Al X-ray source of Kα radiation (1486.6 eV), and all of the spectra had been calibrated with the C 1s peak at 284.6 eV as an internal standard.

3.4. Electrochemical Measurements

The electrochemical properties of the samples were measured on an electrochemical workstation (CHI 760E), through a conventional three-electrode system. Before testing, 5.0 mg of CSs was dispersed in 1.0 mL of a mixed solvent containing anhydrous ethanol (500 μL), H_2O (450 μL) and 5 wt% Nafion (50 μL). The above slurry (4.5 μL) was dropped onto a glassy carbon electrode and used as working electrode. The quality of the commercial 20 wt% Pt/C was half of the CSs catalyst. ORR performance was tested in freshly made KOH aqueous solution (0.1 M) at room temperature. Pt foil and an Ag/AgCl (KCl saturation) electrode were used as the counter electrode and the reference electrode, respectively. The potential in this article was relative to the Ag/AgCl electrode.

4. Conclusions

In summary, a new type of carbon shell co-doped with multi-element including B, N, F and Fe was designed and synthesized. The morphology and the content of the Fe element of the carbon shell could be easily tuned by changing the reaction time between the boronate polymer and Fe^{3+}. The CS_{12-650} catalyst showed excellent catalytic activity toward ORR (E_{onset} = 0.91 V, $E_{half\text{-}wave}$ = 0.82 V vs. RHE), which were comparable to commercial Pt/C in an alkaline system. We verified that the existence of the fluorine element in the carbon matrix was important for the improvement of ORR performance. From the prospective of methodology, we consider that the Kirkendall effect-based route to metallosupramolecular polymer hollow spheres may be of great interest in the design of precursors incorporated with transition metal elements, thereby generating high-performance doped carbon materials.

Supplementary Materials: The following are available online at http://www.mdpi.com/2073-4344/9/1/102/s1, Synthetic routes of DFC and TBB, TEM image of CS_{12-550} and CS_{12-750}, the corresponding K-L plots of CSs, XPS survey spectra CS_{12-550} and CS_{12-750}, the atomic percent of elements of CS_{12-750} and LSV curves of the commercial 20 wt% Pt/C before and after 1000 cycles of CV curves are in the supporting information.

Author Contributions: The experiments were designed by C.Y. and L.D. The experiment was carried out and the manuscript was written by Y.W. The data were analyzed by Y.L., J.M., H.W., T.W., Y.L., B.Z. and Y.X.

Funding: This work was supported by the National Natural Science Foundation of China (51673161, 51773172), Scientific and Technological Innovation Platform of Fujian Province (2014H2006).

Conflicts of Interest: The authors declare no conflict of interest.

References

1. Shao, M.H.; Chang, Q.W.; Dodelet, J.P.; Chenitz, R. Recent Advances in Electrocatalysts for Oxygen Reduction Reaction. *Chem. Rev.* **2016**, *116*, 3594–3657. [CrossRef] [PubMed]

2. Kulkarni, A.; Siahrostami, S.; Patel, A.; Nørskov, J.K. Understanding Catalytic Activity Trends in the Oxygen Reduction Reaction. *Chem. Rev.* **2018**, *118*, 2302–2312. [CrossRef] [PubMed]

3. Mano, N.; Poulpiquet, A.D. O_2 Reduction in Enzymatic Biofuel Cells. *Chem. Rev.* **2018**, *118*, 2392–2468. [CrossRef] [PubMed]

4. Kakati, N.; Maiti, J.; Lee, S.; Jee, S.H.; Viswanathan, B.; Yoon, Y.S. Anode catalysts for direct methanol fuel cells in acidic media: Do we have any alternative for Pt or Pt-Ru? *Chem. Rev.* **2014**, *114*, 12397–12429. [CrossRef] [PubMed]

5. Pegis, M.L.; Wise, C.F.; Martin, D.J.; Mayer, J.M. Oxygen Reduction by Homogeneous Molecular Catalysts and Electrocatalysts. *Chem. Rev.* **2018**, *118*, 2340–2391. [CrossRef] [PubMed]

6. Fu, S.F.; Zhu, C.Z.; Song, J.H.; Engelhard, M.H.; Xia, H.B.; Du, D.; Lin, Y.H. Kinetically Controlled Synthesis of Pt-Based One-Dimensional Hierarchically Porous Nanostructures with Large Mesopores as Highly Efficient ORR Catalysts. *ACS Appl. Mater. Interfaces* **2016**, *8*, 35213–35218. [CrossRef]

7. Unwin, P.R.; Guell, A.G.; Zhang, G.H. Nanoscale Electrochemistry of sp (2) Carbon Materials: From Graphite and Graphene to Carbon Nanotubes. *Acc. Chem. Res.* **2016**, *49*, 2041–2048. [CrossRef]

8. Rao, C.V.; Viswanathan, B. ORR Activity and Direct Ethanol Fuel Cell Performance of Carbon-Supported Pt-M (M = Fe, Co, and Cr) Alloys Prepared by Polyol Reduction Method. *Phys. Chem. C.* **2009**, *113*, 18907–18913.

9. Kim, J.H.; Sa, Y.J.; Jeong, H.Y.; Joo, S.H. Roles of Fe-Nx and Fe-Fe3C@C Species in Fe-N/C Electrocatalysts for Oxygen Reduction Reaction. *ACS Appl. Mater. Interfaces* **2017**, *9*, 9567–9575. [CrossRef]

10. Zhang, Q.; Mamtani, K.; Jain, D.; Ozkan, U.; Asthagiri, A. CO Poisoning Effects on FeNC and CNx ORR Catalysts: A Combined Experimental-Computational Study. *J. Phys. Chem. C* **2016**, *120*, 15173–15184. [CrossRef]

11. Loget, G.; Zigah, D.; Bouffier, L.; Sojic, N.; Kuhn, A. Bipolar electrochemistry: From materials science to motion and beyond. *Acc. Chem. Res.* **2013**, *4*, 2513–2523. [CrossRef] [PubMed]

12. Ambrosi, A.; Chua, C.K.; Bonanni, A.; Pumera, M. Electrochemistry of graphene and related materials. *Chem. Rev.* **2014**, *114*, 7150–7188. [CrossRef]

13. Ghasemi, M.; Daud, W.R.W.; Hassan, S.H.A.; Ismail, O.; Rahimnejad, M.; Jahim, J.M. Nano-structured carbon as electrode material in microbial fuel cells: A comprehensive review. *J. Alloy Compd.* **2013**, *580*, 245–255. [CrossRef]

14. Lin, F.; Liu, Y.J.; Yu, X.Q.; Cheng, L.; Singer, A.; Shpyrko, O.G.; Huolin, L.; Xin, L.H.; Tamura, N.; Tian, C.X.; et al. Synchrotron X-ray Analytical Techniques for Studying Materials Electrochemistry in Rechargeable Batteries. *Chem. Rev.* **2017**, *117*, 13123–13186. [CrossRef] [PubMed]

15. Tong, X.; Wei, Q.L.; Zhan, X.X.; Zhang, G.X.; Shu, H.; Sun, S.H. The New Graphene Family Materials: Synthesis and Applications in Oxygen Reduction Reaction. *Catalysts* **2017**, *7*, 1. [CrossRef]

16. Brouzgou, A.; Song, S.Q.; Liang, Z.X.; Tsiakaras, P. Non-Precious Electrocatalysts for Oxygen Reduction Reaction in Alkaline Media: Latest Achievements on Novel Carbon Materials. *Catalysts* **2016**, *6*, 159. [CrossRef]

17. Wu, Z.X.; Song, M.; Wang, J.; Liu, X. Recent Progress in Nitrogen-Doped Metal-Free Electrocatalysts for Oxygen Reduction Reaction. *Catalysts* **2018**, *8*, 196. [CrossRef]

18. Paraknowitsch, J.P.; Thomasa, A. Doping carbons beyond nitrogen: An overview of advanced heteroatom doped carbons with boron, sulphur and phosphorus for energy applications. *Energy Environ. Sci.* **2013**, *6*, 2839–2855. [CrossRef]

19. Bag, S.; Roy, K.; Gopinath, C.; Raj, C.R. Facile single-step synthesis of nitrogen-doped reduced graphene oxide-Mn(3)O(4) hybrid functional material for the electrocatalytic reduction of oxygen. *ACS Appl. Mater. Interfaces* **2014**, *6*, 2692–2699. [CrossRef] [PubMed]

20. Xiong, D.B.; Li, X.F.; Fan, L.L.; Bai, Z.M. Three-Dimensional Heteroatom-Doped Nanocarbon for Metal-Free Oxygen Reduction Electrocatalysis: A Review. *Catalysts* **2018**, *8*, 301. [CrossRef]

21. Liu, J.; Ji, Y.G.; Qiao, B.; Zhao, F.Q.; Gao, H.X.; Chen, P.; An, Z.W.; Chen, X.B.; Chen, Y. N, S Co-Doped Carbon Nanofibers Derived from Bacterial Cellulose/Poly (Methylene Blue) Hybrids: Efficient Electrocatalyst for Oxygen Reduction Reaction. *Catalysts* **2018**, *8*, 269. [CrossRef]

22. Kim, D.W.; Li, O.L.; Saito, N. Enhancement of ORR catalytic activity by multiple heteroatom-doped carbon materials. *Phys. Chem. Chem. Phys.* **2015**, *17*, 407–413. [CrossRef]

23. Chang, Y.; Yuan, C.H.; Li, Y.T.; Liu, C.; Wu, T.; Zeng, B.R.; Xu, Y.T.; Dai, L.Z. Controllable fabrication of a N and B co-doped carbon shell on the surface of TiO_2 as a support for boosting the electrochemical performances. *J. Mater. Chem. A* **2017**, *5*, 1672–1678. [CrossRef]

24. Chang, Y.; Yuan, C.H.; Liu, C.; Mao, J.; Li, Y.T.; Wu, H.Y.; Wu, Y.Z.; Xu, Y.T.; Dai, L.Z. B, N co-doped carbon from cross-linking induced self-organization of boronate polymer for supercapacitor and oxygen reduction reaction. *J. Power Sources* **2017**, *365*, 354–361. [CrossRef]

25. Bezerra, C.W.B.; Zhang, L.; Lee, K.C.; Liu, H.S.; Marques, A.L.B.; Marques, E.P.; Wang, H.J.; Zhang, J.J. A review of Fe-N/C and Co-N/C catalysts for the oxygen reduction reaction. *Electrochim. Acta* **2008**, *53*, 4937–4951. [CrossRef]

26. Liua, Y.Y.; Yue, X.P.; Li, K.X.; Qiao, J.L.; Wilkinsone, D.; Zhang, J.J. PEM fuel cell electrocatalysts based on transition metal macrocyclic compounds. *Coord. Chem. Rev.* **2016**, *315*, 153–177. [CrossRef]

27. Guo, J.N.; Cheng, Y.H.; Xiang, Z.H. Confined-Space-Assisted Preparation of Fe_3O_4 Nanoparticle Modified Fe-N-C Catalysts Derived from a Covalent Organic Polymer for Oxygen Reduction. *ACS Sustain. Chem. Eng.* **2017**, *5*, 7871–7877. [CrossRef]

28. Liu, J.; Li, E.L.; Ruan, M.B.; Song, P.; Xu, W.L. Recent Progress on Fe/N/C Electrocatalysts for the Oxygen Reduction Reaction in Fuel Cells. *Catalysts* **2015**, *5*, 1167–1192. [CrossRef]

29. Jiang, W.J.; Gu, L.; Li, L.; Zhang, Y.; Zhang, X.; Zhang, L.J.; Wang, J.Q.; Hu, J.S.; Wei, Z.D.; Wan, L.J. Understanding the High Activity of Fe-N-C Electrocatalysts in Oxygen Reduction: Fe/Fe_3C Nanoparticles Boost the Activity of Fe-N(x). *J. Am. Chem. Soc.* **2016**, *138*, 3570–3578. [CrossRef]

30. Li, L.Y.; Yuan, C.H.; Dai, L.Z. Thermoresponsive Polymeric Nanoparticles: Nucleation from Cooperative Polymerization Driven by Dative Bonds. *Macromolecules* **2014**, *47*, 5869–5876. [CrossRef]

31. Li, L.Y.; Yuan, C.Y.; Zhou, D.M.; Ribbe, D.E.; Kittilstved, P.R.; Thayumanavan, S. Utilizing Reversible Interactions in Polymeric Nanoparticles to Generate Hollow Metal-Organic Nanoparticles. *Angew. Chem. Int. Ed.* **2015**, *127*, 13183–13187. [CrossRef]

32. Ejima, H.; Richardson, J.J.; Liang, K.; Best, J.P.; Koeverden, M.P.; Such, G.H.; Cui, J.W.; Such, G.H.; Cui, J.W.; Caruso, F. One-step assembly of coordination complexes for versatile film and particle engineering. *Science* **2013**, *341*, 154–157. [CrossRef]

33. Rahim, M.A.; Bjornmalm, M.; Suma, T.; Faria, M.; Ju, Y.; Kempe, K.; Mgllner, M.; Ejima, H.; Stickland, A.D.; Caruso, F. Metal-Phenolic Supramolecular Gelation. *Angew. Chem. Int. Ed.* **2016**, *55*, 13803–13807. [CrossRef]

34. Barrett, D.G.; Fullenkamp, D.E.; He, L.H.; Holten, N.; Lee, K.C. pH-Based Regulation of Hydrogel Mechanical Properties Through Mussel-Inspired Chemistry and Processing. *Adv. Funct. Mater.* **2013**, *23*, 1111–1119. [CrossRef]

35. He, J.R.; Luo, L.; Chen, Y.F.; Manthiram, A. Yolk-Shelled C@Fe$_3$O$_4$ Nanoboxes as Effcient Sulfur Hosts for High-Performance Lithium-Sulfur Batteries. *Adv. Mater.* **2017**, *29*, 1702707. [CrossRef] [PubMed]

36. Liu, X.X.; Wang, Y.H.; Dong, L.; Chen, X.; Xin, G.X.; Zhang, Y.; Zang, J.B. One-step synthesis of shell/core structural boron and nitrogen co-doped graphitic carbon/nanodiamond as efficient electrocatalyst for the oxygen reduction reaction in alkaline media. *Electrochim. Acta* **2016**, *194*, 161–167. [CrossRef]

37. Sun, X.J.; Zhang, Y.W.; Song, P.S.; Pan, J.; Zhuang, L.; Xu, W.L.; Xing, W. Fluorine-Doped Carbon Blacks: Highly Efficient Metal-Free Electrocatalysts for Oxygen Reduction Reaction. *ACS Catal.* **2013**, *3*, 1726–1729. [CrossRef]

38. Ali-Löytty, H.; Louie, M.W.; Singh, M.R.; Li, L.; Casalongue, H.G.; Ogasawara, H.; Crumlin, E.J.; Liu, Z.; Bell, A.T.; Nilsson, A.; et al. Ambient-Pressure XPS Study of a Ni-Fe Electrocatalyst for the Oxygen Evolution Reaction. *J. Phys. Chem. C* **2016**, *120*, 2247–2253. [CrossRef]

39. Wang, H.Y.; Iyyamperumal, E.; Roy, A.; Xue, Y.H.; Yu, D.S.; Dai, L.M. Vertically aligned BCN nanotubes as efficient metal-free electrocatalysts for the oxygen reduction reaction: A synergetic effect by co-doping with boron and nitrogen. *Angew. Chem. Int. Ed.* **2011**, *50*, 11756–11760. [CrossRef]

Article

Electro-Reduction of Molecular Oxygen Mediated by a Cobalt(II)octaethylporphyrin System onto Oxidized Glassy Carbon/Oxidized Graphene Substrate

Camila Canales [1], Leyla Gidi [1], Roxana Arce [2], Francisco Armijo [1], María J. Aguirre [3],* and Galo Ramírez [1],*

1 Departamento de Química Inorgánica, Facultad de Química y de Farmacia, Pontificia Universidad Católica de Chile, Av. Vicuña Mackenna 4860, Casilla 306, Correo 22, Santiago 7820436, Chile; cicanale@uc.cl (C.C.); ldgidi@uc.cl (L.G.); jarmijom@uc.cl (F.A.)
2 Departamento Ciencias Biológicas y Químicas, Facultad de Medicina y Ciencia, Universidad San Sebastián, Lota 2465, Providencia, Santiago 7510157, Chile; arce.roxana@gmail.com
3 Departamento de Química de los Materiales, Facultad de Química y Biología, Universidad de Santiago de Chile USACH. Av. L.B. O'Higgins 3363, Santiago 9170019, Chile
* Correspondence: maria.aguire@usach.cl (M.J.A.); gramirezj@uc.cl (G.R.)

Received: 5 November 2018; Accepted: 3 December 2018; Published: 6 December 2018

Abstract: The oxygen reduction reaction (ORR) is the most important reaction in life processes and in energy transformation. The following work presents the design of a new electrode which is composed by deposited cobalt octaethylporphyrin onto glassy carbon and graphene, where both carbonaceous materials have been electrochemically oxidized prior to the porphyrin deposition. The novel generated system is stable and has an electrocatalytic effect towards the oxygen reduction reaction, as a result of the significant overpotential shift in comparison to the unmodified electrode and to the electrodes used as target. Kinetic studies corroborate that the system is capable of reducing molecular oxygen via four electrons, with a Tafel slope value of 60 mV per decade. The systems were morphologically characterized by scanning electron microscopy (SEM) and atomic force microscopy (AFM) Electrochemical impedance spectroscopy studies showed that the electrode previously oxidized and modified with cobalt porphyrin is the system that possesses lower resistance to charge transfer and higher capacitance.

Keywords: oxygen reduction reaction; glassy carbon electrode; graphene; metalloporphyrins

---

1. Introduction

The oxygen reduction reaction (ORR) is determinant for the operation of devices associated with the conversion and storage of energy, such as fuel cells and metal–air batteries, among others [1–5]. In this sense, the development of electrocatalysts for ORR becomes crucial both for the use and commercialization of these technologies [6]. Nowadays, the best electrocatalysts for ORR are associated with high cost materials, related to Pt and derived materials, such as Pt nanoparticles onto other low-cost substrates [7–10]. However, the scarcity of this material, as well as the high cost associated with it, means that large-scale applications are very difficult [11,12]. On the other hand, the feasibility of generating low-cost electrocatalysts to reduce molecular oxygen at low overpotentials has been demonstrated. Consequently, this has proved that carbon-based materials, such as activated carbon (carbon black), carbon nanotubes, carbon nanofibers, and graphene, among others, play an important role in overcoming challenges concerning technology, energy, and practical applications. In this regard, these materials have been widely used and are promising in applications of great importance, such as in the generation of energy via electrochemical and electrocatalytic processes [13]. Therefore, carbonaceous materials, such as glassy carbon (GC) and graphene (GPH), can be excellent choices to be used as electrode substrates [14]. Thus, since graphene-based materials emerged, they have been identified as interesting candidates to be applied as catalysts due to their high conductivity, large surface area, high chemical/electrochemical stability, as well as their strong adherence to compounds with catalytic characteristics [15–17]. In addition, the large quantities of functional groups that an oxidized graphene may contain can provide more active sites for nucleation processes and coordinate catalytic compounds [18]. Thus, several works based on carbonaceous materials have been published, which demonstrate the effectiveness of these systems for multiple reactions of environmental and energy interest [19–23].

On the other hand, several studies are focused on substituting the use of Pt as an electrocatalyst material. In addition, carbonaceous materials allow to reduce costs compared to Pt-based materials, where both GC and GPH have great advantages in terms of cost. Nevertheless, the use of metal complexes (non-noble) has been used in order to generate new electrode systems that are capable of reducing molecular oxygen (e.g., Fe, Ni, Co), where Co(II) complexes have attracted great attention. Among these complexes, aza-macrocyclic type compounds are found, such as phthalocyanines and porphyrins, which possess excellent redox properties [24,25] that make them candidates to modify electrode substrates so as to electrocatalyze multiple reactions, such as oxygen reduction reaction, hydrogen evolution reaction, and carbon dioxide reduction, which are important reactions in the area of energy generation and storage. Additionally, multiple electrode systems have been generated using this type of complexes [26–29], which allows the obtainment of significant potential shifts and high efficiencies towards ORR. Previously, our group reported a very electrocatalytic system, which was capable of reducing molecular oxygen via four electrons using a commercial octaethylated cobalt porphyrin onto oxidized GC. In that study, it was found that oxidized groups serve as anchoring molecules that permit the coordination of a first layer of metalloporphyrin and provide stability to the system [30]. After the coordination of the first layer, the porphyrins are then arranged in a columnar side-to-side form over the covalently modified system, as demonstrated by RAMAN spectroscopy, X-ray diffraction (XRD), and UV-Vis spectroscopy [31,32].

Finally, considering the abovementioned background, this paper presents the generation of a simple and active GC/GPH/Co(II)EPO system (Figure 1), where carbon materials are previously oxidized at a fixed potential (+1.6 V versus Ag/AgCl and +1.0 versus Ag/AgCl for GC and GPH, respectively). Oxidized groups are generated on the surface in order to serve as anchoring molecules to coordinate the first layer of cobalt porphyrins. In this way, it is expected that the system could aid as an electrocatalyst towards the molecular oxygen reduction reaction as a result of the contribution of glassy carbon, graphene, oxidized groups, and cobalt porphyrin, in a synergistic way. Concomitantly, the existence of these anchoring molecules is expected to stabilize the generated system, so as to have stable and reproducible responses over time.

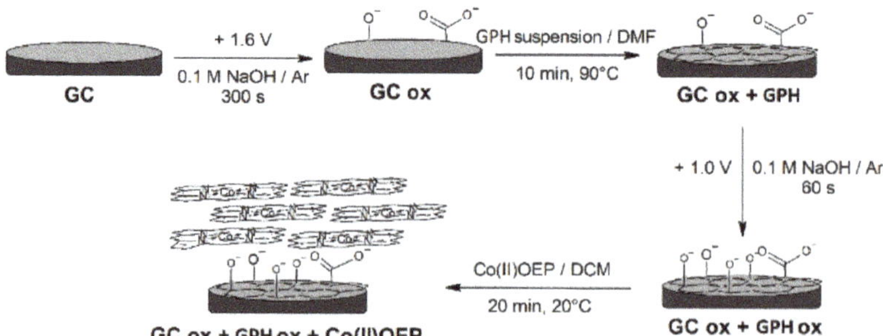

Figure 1. Modification diagram of vitreous carbon electrode (GC = Glassy Carbon, ox = oxidized, GPH = Graphene, Co(II)OEP = cobalt(II)octaethylporphine, DFM = dimethylformamide, DCM = dichloromethane).

2. Results and Discussion

2.1. Morphological Studies of Modified Systems

In order to study the systems in terms of morphology, SEM studies were conducted for the GC, GC + GPH, GC ox + GPH ox, and GC ox + GPH ox + Co(II)EPO systems. Figure 2 shows the images corresponding to these studies. As observed, noticeable differences exist in comparison to the surface morphologies presented by each system.

The addition of GPH onto the GC substrate causes the dispersion of GPH agglomerations along the entire surface of the electrode. However, when the GPH is deposited on previously oxidized GC (GC ox + GPH), and also when the GC ox + GPH ox system is subsequently generated, many more covered substrates are obtained, with larger agglomerations than those observed by the non-oxidized GC + GPH system.

Finally, it is possible to observe how the metallic cobalt porphyrin deposits change the morphology of the electrode system, owing to the appearance of fibers on the oxidized surface. It is noteworthy that these porphyrin deposits are mostly arranged on the agglomerations rather than on the unmodified surface, which corroborates that the oxidized groups help the metal complex to be coordinated to the system more easily. On the other hand, AFM images, shown next to each SEM image, complement the morphological analysis of the resulting surfaces in terms of roughness (R_q) variation. As can be seen, the initial GC substrate has a R_q value of 6.0 nm, being the lowest in comparison to the modified systems whose R_q have values of up to 355.0 nm (Table 1). In this regard, the modified system GC ox + GPH ox + Co(II)OEP presents the highest roughness value in comparison to the rest of the systems, which would explain the increment in both the capacitance of the system and the total current, as will be seen later. At the same time, the formation of agglomerations is observed, which may indicate the formation of columnar systems given the nature of porphyrin complexes that have the ability to be arranged one over another through π-stacking [30].

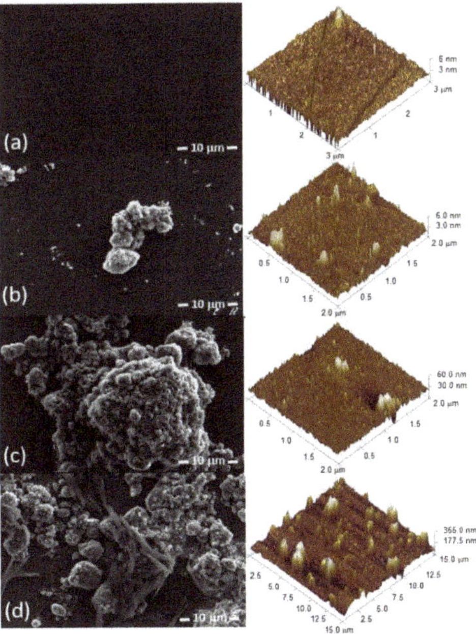

Figure 2. Morphological studies by SEM (**left**) and AFM (**right**) for (**a**) GC, (**b**) GC + GPH, (**c**) GC ox + GPH ox and (**d**) GC ox + GPH ox + Co(II)OEP systems.

Table 1. R_q values for the GC systems and modified GC systems.

System	R_q (nm)
GC	6.0
GC + GPH	6.0
GC ox + GPH ox	60.0
GC ox + GPH ox + Co(II)OEP	355.0

2.2. Electrochemical Response of the Modified Systems

Figure 3a shows the voltammetric responses of GC, GC + GPH, GC + GPH + Co(II)EPO, GC ox + GPH ox, and GC ox + GPH ox + Co(II)EPO systems. As shown in Figure 3a, all the modified systems have a different voltammetric profile and potential shifts compared to the initial substrate. In this regard, a significant shift on the onset potential of 400 mV appears compared to the GC, associated to the GC ox + GPH ox + Co(II)OEP system, which, in turn, presents a significant current increment, which reaches a cathodic peak current of 80 μA. Consequently, it is possible to demonstrate the synergistic effect between the oxidized carbonaceous materials (GC ox + GPH ox) and the porphyrin cobalt complex (Co(II)OEP), since, in the first case, the potential shift of the onset potential is small in comparison to the ones shown by the initial substrates (GC + GPH). However, after the deposition of the cobalt complex, both GC + GPH + Co(II)OEP and GC ox + GPH ox + Co(II)OEP systems present significant onset potential shifts, due to the contribution of the metallic complex, as has been demonstrated in previous works with similar systems [30,33–35]. This phenomenon is typical in electrode systems that are modified with cobalt-containing compounds, such as cobalt oxide (II, III), Co_3O_4 [36,37]. It is remarkable that the porphyrin complex deposited on the GC modified with non-oxidized GPH is substantially less catalytic than the system shown here, where both carbonaceous substrates have been previously oxidized. This result draws attention because graphene oxidation should bring about a conductivity loss associated with the loss of sp^2 hybridization that is not observed in the results here obtained. By contrast, oxidized carbonaceous and non-oxidized materials show a similar profile for the oxygen reduction, with the same onset potentials. In this sense, it is important to highlight this latter effect: The manifested synergy between the Co(II)OEP and carbon-based materials is much greater when the carbonaceous materials are previously oxidized.

Thus, the porphyrin-modified oxidized system has the greatest electrocatalytic effect towards ORR, reducing molecular oxygen to an onset potential of −0.08 V.

As well as the faradaic current increases in the GC ox + GPH ox + Co(II)OEP system, the capacitive current also presents a high increment in comparison to the bare GC (Figure 3b).

The existence of oxidized groups that serve as covalent anchoring groups to the first layer of deposited porphyrins generates an increment in the number of components on the electrode surface, which results in the double layer increment (capacitance) and, therefore, in the capacitive current presented by the modified system [38].

On the other hand, as seen in Figure 3b, the modified system shows a slight anodic signal at ca. 0.0 V, associated to the Co(II)/Co(III) redox couple. Thus, the generation of μ-oxo bonding is formed by the porphyrin central metal, which implies the formation of oxidized cobalt species that facilitates, finally, the molecular oxygen reduction [39].

By contrast, the existence of oxidized graphene provokes a synergistic effect on the electrodic system, by increasing the electrocatalytic effect that the GC ox + Co(II)OEP system already possesses, as reported in a previous work of our group [30]. Thus, the existence of oxidized graphene (GPH ox) is crucial to achieve a greater electrocatalytic effect and the addition of robustness to the modified electrode.

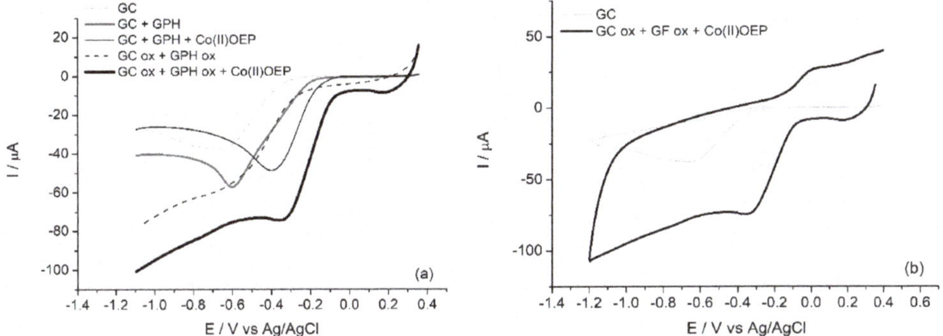

Figure 3. (**a**) Voltammetric profiles of all studied systems, (**b**) cyclic voltammetry profiles for GC and GC ox + GPH ox + Co(II)OEP systems, in 0.1 M NaOH saturated with O_2. $v = 0.1$ Vs^{-1}.

2.3. Electrochemical Kinetics

In order to know how ORR occurs at the electrode–solution interphase, the corresponding kinetic study of the GCox + GPH ox + Co(II) OEP system was conducted. Figure 4a shows the linearity of the logarithm of the peak current ($\log I_p$) vs. logarithm of the scan rate ($\log v$), which gives a slope value of 0.48 (~0.5) that implies that the ORR process is diffusion-controlled [40]. Figure 4b depicts the variation of the peak current vs. the square root of the scan rate, which affords a linear plot ($R^2 = 0.998$); these results are consistent with the diffusion-controlled process.

This background enables the determination of the number of transferred electrons, n, during the reduction reaction using the Randles–Sevcik equation (Equation (1)) for irreversible diffusion-controlled systems, such as the one shown in this case [40,41]. In this equation, α is the charge transfer coefficient, n_a the number of electrons involved in the rate-determining step of the reaction, D_o (2×10^{-5} cm^2 s^{-1}) [42] the diffusion coefficient of the electroactive specie, and C$_o$, the concentration of O_2 on the solution, which corresponds to 6.90×10^{-7} mol cm^{-3} in O_2 saturated aqueous solutions at room temperature [43].

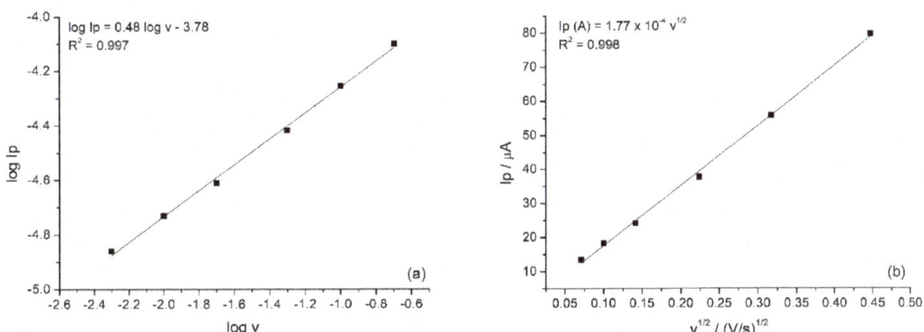

Figure 4. Kinetic studies. (**a**) Dependence of the logarithm of the peak current on scan rate logarithm. (**b**) Dependence of peak current on square-root of scan rate.

Then, considering the information given in Figure 4 and the Randles–Sevcik equation (Equation (1)) described above, it is possible to infer that:

$$I_p = (2.99 \times 10^5)\, n\, [(1 - \alpha)n_a]^{1/2}\, C_o\, A\, D_o^{1/2}\, v^{1/2} \tag{1}$$

$$\frac{I_P}{v^{1/2}} = 177 \times 10^{-6}\, \mathrm{A}\left(\mathrm{Vs}^{-1}\right)^{-1/2} \tag{2}$$

$$\frac{I_P}{v^{1/2}} = \left(2.99 \times 10^{5}\right) n\, [(1-\alpha)n_a]^{1/2}\, 6.90 \times 10^{-7}\, \mathrm{mol\ cm}^{-3}\, 0.070\ \mathrm{cm}^2 \left(2 \times 10^{-5}\ \mathrm{cm}^2\,\mathrm{s}^{-1}\right)^{1/2} \tag{3}$$

$$177 \times 10^{-6}\, \mathrm{A}\left(\mathrm{Vs}^{-1}\right)^{-1/2} = \left(6.46 \times 10^{-5}\right) n\, [(1-\alpha)n_a]^{1/2} \tag{4}$$

$$2.7 = n\, [(1-\alpha)n_a]^{1/2} \tag{5}$$

Furthermore, considering that $(1-\alpha)n_a$ must be known, an equation based on its dependence on the difference between the peak potential (E_p) and the half-peak potential ($E_{p/2}$) was employed as follows [44]:

$$(1-\alpha)n_a = 0.0477\ \mathrm{V}\,/\,|(-0.332\ \mathrm{V} + 0.184\ \mathrm{V})| = 0.32 \tag{6}$$

$$[(1-\alpha)n_a]^{1/2} = 0.6 \tag{7}$$

$$n = \frac{2.7}{0.6} = 4.6\ \mathrm{electrons} \tag{8}$$

Finally, the trend of the system to reduce the electroactive species via four electrons is demonstrated.

Concomitantly, the Tafel slope for the proposed system was calculated. Figure 5 shows the obtained plot at low overpotentials, which gives a slope value of $2.303RT/\alpha F \approx 60$ mV per decade that corresponds to a chemical step subsequent to the first electronic transfer, as the rate-determining step of the reaction (Rds) [45]. In this sense, there would be a direct oxygen reduction reaction involving four electrons, where the first electronic transfer is followed by the breaking of the double O–O bond, which would comprise the slow step, without the formation of peroxide intermediates [46–48]. Consequently, the following mechanism may be proposed:

$$\begin{aligned}
\mathrm{Co(II)\text{-}O_2} + e^- &\rightarrow \mathrm{Co(II)\text{-}O_2^-} \\
\mathrm{Co(II)\text{-}O_2^-} + e^- &\rightarrow \mathrm{Co(II)\text{-}O_2^{2-}} \\
\mathrm{Co(II)\text{-}O_2^{2-}} + e^- &\rightarrow \mathrm{Co(II)\text{-}O_2^{3-}} \\
\mathrm{Co(II)\text{-}O_2^{3-}} + e^- &\rightarrow \mathrm{Co(II)\text{-}O_2^{4-}}
\end{aligned}
\qquad \text{Consecutive stages}$$

$$\mathrm{Co(II)\text{-}O_2^{4-}} + 2H_2O \rightarrow \mathrm{Co(II)} + 4OH^- \qquad \longrightarrow \quad \text{Rate determining step}$$

This mechanism includes the previous stage of an adduct formation between the metallic center and the oxygen, adduct where internal charge transfer from the metallic center to the oxygen can take place, which is not included in the above-shown mechanism in order to simplify the steps. In the four-electron transfer, the Co metal center behaves as an electronic bridge that would transfer the charge towards the oxygen bond, until this molecule receives four electrons. These first four stages would occur consecutively, at very close potentials, which give rise to a single, rather wide, cathodic wave (Figure 3b) due to the small differences in energy that each electron transfer requires. At that point, considering these four fast steps and close in energy, the rate determining step would comprise the reaction of the negative charged adduct with water from the medium to break up, generating hydroxyl ions as final products.

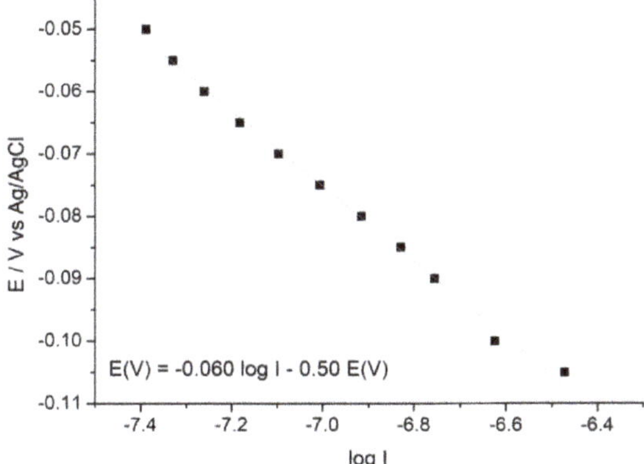

Figure 5. Potential dependence on current logarithm (Tafel slope), for the GC ox + GPH ox + Co(II)OEP system.

Finally, in order to corroborate the stability of the generated system, chronoamperometry was performed at fixed potential during 3600 s. As appreciated in Figure 6, the system attains a current response that remains constant over time, evidencing that the system, as well as being efficient and easy to obtain, is stable.

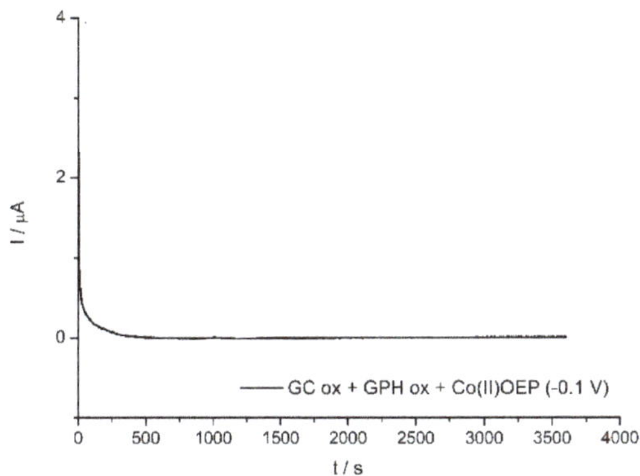

Figure 6. Chronoamperometric study of the GC ox + GPH ox + Co(II)OEP system stability.

2.4. Electrochemical Impedance Spectroscopy

Figure 7 shows the Nyquist diagram for the GC, GC ox + GPH ox, and GC ox + GPH ox + Co(II)OEP systems, obtained at a fixed potential of −0.4 V. The equivalent circuit, whose parameters are summarized in Table 2, is also shown. Rs is the solution resistance, while Rct and CPE are the charge transfer resistance and the constant phase element, respectively.

As is observed, both modified systems, GC ox + GPH ox and GC ox + GPH ox + Co(II)OEP, show a great decrease of charge transfer resistance, graphically associated to the semicircle closure. The GC ox + GPH ox + Co(II)OEP system is the one with the lowest charge transfer resistance, which is consistent with that obtained in the voltammetric studies, which were previously accomplished.

In the equivalent circuit, a constant phase element (CPE) associated with two parameters (T and P) was used. It is assumed that if P is close to 1, T corresponds to the capacitance. If P is close to 0.5, T is associated with diffusion species, whereas if P is close to 0, T will be a resistance value [49]. Table 2 shows the results obtained for this experiment and it can be observed that, in all cases, the parameter associated with P is close to 1, indicating that T would yield capacitance values, which are higher when rusted systems are used. In addition, it is noteworthy that for oxidized systems, the capacitance value is similar before and after the porphyrin addition. Consequently, it was corroborated that the final obtained capacitance is highly influenced by the oxidized groups in the GC and GPH, respectively, rather than by the presence of the porphyrin complex. Additionally, it can be observed that the diminution of the Rct value is consistent with the electrocatalytic activity, which is given by the overpotential onset value. Concomitantly, the system obtained by GC ox + GPH ox + Co(II)OEP is the most conductive, which evidences its electrocatalytic effect towards ORR.

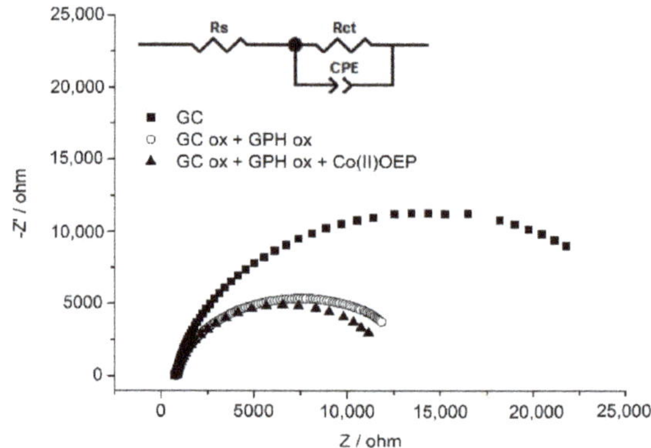

Figure 7. Nyquist plot and equivalent circuit used for GC, GC ox + GPH ox, and GC ox + GPH ox + Co(II)OEP systems, in O_2-saturated 0.1 M NaOH solution at E = -0.4 V, frequency range 0.1–100,000 Hz.

Table 2. Parameters obtained with the equivalent circuit of Figure 7. E_O is the onset potential, Rs is solution resistance, Rct the charge transfer resistance, and CPE the constant phase element.

System	E_O (V)	Rs (Ω)	Rct (Ω)	CPE (T, P) (s Ω^{-1})
GC	-0.34	742.1	27,244	3.53×10^{-6}, 0.88
GC ox + GPH ox	-0.25	810.9	12,999	5.20×10^{-6}, 0.88
GC ox + GPH ox + Co(II)OEP	-0.08	796.1	11,905	5.22×10^{-6}, 0.88

Finally, Table 3 compares the generated material (GC ox + GPH ox + Co(II)OEP) with other similar systems that have been previously reported in the literature towards the ORR. In this sense, the electrode here developed is found to be one of the most electroactive systems due to its onset overpotential. Moreover, this new material is easily achieved, highly stable, and cheap.

Table 3. Comparison between similar systems towards the oxygen reduction reaction (ORR).

Material	Medium	ORR Onset (V)	Number Electrons ORR	Reference
GC/GPH	0.1 M KOH	-0.18 vs. SCE	$\approx$4	[50]
Cu foil/GPH	0.1 M KOH	-0.3 vs. Ag/AgCl	$\approx$2	[51]
GC/GPH Quantum Dots	0.1 M KOH	-0.16 vs. Ag/AgCl	3.6–4.4	[52]
CoTPyP/PSS-rGO *	0.1 M KOH	-0.15 vs. Ag/AgCl	3.61–3.67	[53]
CoP-CMP800 **	0.1 M KOH	-0.1 vs. Ag/AgCl	3.83–3.86	[26]
GC Co(II)OEP	0.1 M NaOH	-0.2 vs. Ag/AgCl	-	[54]
GC Co(II)OEP-Fe(III)OEP	0.1 M NaOH	-0.13 vs. Ag/AgCl	$\approx$4	[54]
GC ox/Co(II)OEP	0.1 M NaOH	$+0.05$ vs. Ag/AgCl	$\approx$4	[30]
GC ox + GPH ox + Co(II)OEP	0.1 M NaOH	-0.08 vs. AgAgCl	$\approx$4	This work

* Cobalt porphyrin on poly(sodium-pstyrenesulfonate) modified reduced graphene oxide. ** Cobalt porphyrin-based conjugated mesoporous polymers.

3. Materials and Methods

3.1. Equipment

Cyclic Voltammetry (CV), chronoamperometry, and electrochemical impedance spectroscopy (EIS) studies were accomplished on a CH Instrument 750D potentiostat galvanostat (Austin, USA). A conventional three-electrode system consisting of a glassy carbon working electrode, a large area platinum wire counter electrode, and Ag/AgCl (3 M, KCl) reference electrode was used for all the measurements. All potentials quoted in this work are relative to this reference electrode.

3.2. Reagents

Potassium chloride, sodium hydroxide, and graphene nanoplatelets were purchased from Merck (Darmstadt, Germany). De-ionized water was obtained from a Millipore-Q system (Darmstadt, Germany) (18.2 MΩ·cm). Argon (99.99% pure) and dioxygen gas (99.99% pure) were purchased from Linde Chile (Santiago, Chile). Cobalt porphyrin (2,3,7,8,12,13,17,18-octaethyl-21H,23H-porphine cobalt(II)) as well as dichloromethane (DCM, HPLC grade) and dimethylformamide (99.90% DMF) were purchased from J.T. Baker (USA).

3.3. Preparation of Modified Electrodes

The glassy carbon electrode was polished to a mirror finish on a felt pad using alumina slurries (3 μm), sonicated for 120 s to remove the excess of alumina and finally, cycling the potential between -0.7 V and 0.7 V in a 0.1 M NaOH solution (Ar atmosphere) until stabilization of the electrode response (20 cycles).

To obtain the GC + GPH system, a drop of 0.5 mg·mL^{-1} of graphene suspension (5 mg in 10 mL of dimethylformamide) was placed onto the clean (not oxidized) GC surface, and dried at 95 °C for 10 min to evaporate the excess of dimethylformamide (DMF).

The oxidized GC (GC ox) was obtained by applying a constant anodic potential (+1.6 V) until the system accumulated a charge of 0.1 C (300 s approximately) in 0.1 M NaOH. Then, one drop of the graphene suspension was placed upon the GC ox surface and dried as aforementioned. After that procedure, a constant anodic potential (+1.0 V) was again applied for 60 s, in order to generate the GC ox + GPH ox system. Finally, the obtained system was immersed into a metalloporphyrin solution, M–OEP (M = Co(II)) 0.2 mM dissolved in CH_2Cl_2 for 20 min and finally dried at room temperature for 1 minute (GC ox + GPH ox + Co(II)OEP). To obtain the GC + GPH + Co(II)OEP blank system, the oxidation procedure was omitted. The electrodes were rinsed with fresh deionized water after each modification step.

3.4. Instrumentation

All modified electrodes were analyzed using CV in a 0.1 M NaOH solution saturated with dioxygen (bubbling during 20 min before each measurement) by cycling the potential from 0.4 to -1.1 V.

Chronoamperometry was conducted in a 0.1 M NaOH solution saturated with dioxygen. The modified electrode was tested by measuring the current at a constant potential of -0.1 V for 1000 s.

Electrochemical impedance spectroscopy (EIS) was performed at a constant potential of -0.4 V in a 0.1 M NaOH solution saturated with dioxygen at frequencies that range between 0.1 and 100,000 Hz.

The scanning electron microscopy studies were carried out on a SEM LEO 1420 VP equipment (Cambridge, England).

4. Conclusions

In the current work, a new system consisting of cobalt octaethylporphyrin on glassy carbon and graphene was obtained; both carbonaceous materials have been previously electrochemically oxidized to carry out the porphyrin deposition of (GC ox + GPH ox + Co(II)OEP). In addition to being stable, this system exhibited the ability to electrocatalyze the ORR via four electrons, showing a large potential shift towards positive values in relation to independent substrates, unoxidized or oxidized, and without porphyrin deposit. This new material presents a Tafel slope of 60 mV per decade, indicating that is a fast kinetic system, where the rate determining step corresponds to a chemical one. The studies conducted using electrochemical impedance showed that this novel system possesses lower charge transfer resistance, as well as a higher capacitance compared to glassy carbon and oxidized carbonaceous substrates.

Author Contributions: C.C. and L.G. performed the experiments. R.A. and F.A. investigated the relevant literature and helped to write this manuscript. M.J.A. and G.R. wrote, reviewed, and modified this manuscript. All authors contributed to the general discussion about this work.

Funding: This research received no external funding.

Acknowledgments: This work was supported by FONDECYT Project No.: 1170340; FONDECYT Project N° 1160324; Dicyt-Usach, and Doctoral Scholarship CONICYT Project No.: 21170542.

Conflicts of Interest: The authors declare no conflict of interest.

References

1. Jiang, W.J.; Gu, L.; Li, L.; Zhang, Y.; Zhang, X.; Zhang, L.J.; Wang, J.Q.; Hu, J.S.; Wei, Z.; Wan, L.J. Understanding the high activity of Fe-N-C electrocatalysts in oxygen reduction: Fe/Fe$_3$C nanoparticles boost the activity of Fe-N(x). *J. Am. Chem. Soc.* **2016**, *138*, 3570–3578. [CrossRef] [PubMed]

2. Geng, D.; Chen, Y.; Chen, Y.; Li, Y.; Li, R.; Sun, X.; Ye, S.; Knights, S. High oxygen-reduction activity and durability of nitrogen-doped graphene. *Energy Environ. Sci.* **2011**, *4*, 760–764. [CrossRef]

3. Steele, B.C.; Heinzel, A. Materials for fuel-cell technologies. *Nature* **2001**, *414*, 345–352. [CrossRef] [PubMed]

4. Liang, H.W.; Zhuang, X.; Bruller, S.; Feng, X.; Mullen, K. Hierarchically porous carbons with optimized nitrogen doping as highly active electrocatalysts for oxygen reduction. *Nat. Commun.* **2014**, *5*, 4973. [CrossRef]

5. Van der Vliet, D.F.; Wang, C.; Tripkovic, D.; Strmcnik, D.; Zhang, X.F.; Debe, M.K.; Atanasoski, R.T.; Markovic, N.M.; Stamenkovic, V.R. Mesostructured thin films as electrocatalysts with tunable composition and surface morphology. *Nat. Mater.* **2012**, *11*, 1051–1058. [CrossRef] [PubMed]

6. Zhou, T.; Ma, R.; Zhou, Y.; Xing, R.; Liu, Q.; Zhu, Y.; Wang, J. Efficient N-doping of hollow core-mesoporous shelled carbon spheres via hydrothermal treatment in ammonia solution for the electrocatalytic oxygen reduction reaction. *Microporous Mesoporous Mater.* **2018**, *261*, 88–97. [CrossRef]

7. Alia, S.M.; Jensen, K.; Contreras, C.; Garzon, F.; Pivovar, B.; Yan, Y. Platinum coated copper nanowires and platinum nanotubes as oxygen reduction electrocatalysts. *ACS Catal.* **2013**, *3*, 358–362. [CrossRef]

8. Narayanamoorthy, B.; Datta, K.K.R.; Eswaramoorthy, M.; Balaji, S. Improved oxygen reduction reaction catalyzed by Pt/Clay/nafion nanocomposite for PEM fuel cells. *ACS Appl. Mater. Interfaces* **2012**, *4*, 3620–3626. [CrossRef]

9. Zhang, Y.; Han, T.; Fang, J.; Xu, P.; Li, X.; Xu, J.; Liu, C.-C. Integrated Pt$_2$Ni alloy@Pt core-shell nanoarchitectures with high electrocatalytic activity for oxygen reduction reaction. *J. Mater. Chem. A* **2014**, *2*, 11400–11407. [CrossRef]

10. Zhao, R.; Liu, Y.; Liu, C.; Xu, G.; Chen, Y.; Tang, Y.; Lu, T. Pd@Pt core-shell tetrapods as highly active and stable electrocatalysts for the oxygen reduction reaction. *J. Mater. Chem. A* **2014**, *2*, 20855–20860. [CrossRef]

11. Gasteiger, H.A.; Kocha, S.S.; Sompalli, B.; Wagner, F.T. Activity benchmarks and requirements for Pt, Pt-alloy, and non-Pt oxygen reduction catalysts for PEMFCs. *Appl. Catal. B Environ.* **2005**, *56*, 9–35. [CrossRef]

12. Chen, A.; Holt-Hindle, P. Platinum-based nanostructured materials: Synthesis, properties, and applications. *Chem. Rev.* **2010**, *110*, 3767–3804. [CrossRef] [PubMed]

13. You, P.Y.; Kamarudin, S.K. Recent progress of carbonaceous materials in fuel cell applications: An overview. *Chem. Eng. J.* **2017**, *309*, 489–502. [CrossRef]

14. Chabot, V.; Higgins, D.; Yu, A.; Xiao, X.; Chena, Z.; Zhang, J. A review of graphene and graphene oxide sponge: Material synthesis and applications to energy and the environment. *Energy Environ. Sci.* **2014**, *7*, 1564–1596. [CrossRef]

15. Liang, Y.; Li, Y.; Wang, H.; Zhou, J.; Wang, J.; Regier, T.; Dai, H. Co$_3$O$_4$ nanocrystals on graphene as a synergistic catalyst for oxygen reduction reaction. *Nat. Mater.* **2011**, *10*, 780–786. [CrossRef] [PubMed]

16. Zhou, W.; Ge, L.; Chen, Z.-G.; Liang, F.; Xu, H.-Y.; Motuzas, J.; Julbe, A.; Zhu, Z. Amorphous iron oxide decorated 3D heterostructured electrode for highly efficient oxygen reduction. *Chem. Mater.* **2011**, *23*, 4193–4198. [CrossRef]

17. Wu, J.; Zhang, D.; Wang, Y.; Wan, Y.; Hou, B. Catalytic activity of graphene–cobalt hydroxide composite for oxygen reduction reaction in alkaline media. *J. Power Sources* **2012**, *198*, 122–126. [CrossRef]

18. Bai, H.; Li, C.; Shi, G. Functional composite materials based on chemically converted graphene. *Adv. Mater.* **2011**, *23.9*, 1089–1115. [CrossRef]

19. Alenezi, K. Electrocatalytic Hydrogen Evolution Reaction Using mesotetrakis-(pentafluorophenyl)porphyrin iron(III) chloride. *Int. J. Electrochem. Sci.* **2017**, *12*, 812–818. [CrossRef]

20. Guo, D.; Shibuya, R.; Akiba, C.; Saji, S.; Kondo, T.; Nakamura, J. Active sites of nitrogen-doped carbon materials for oxygen reduction reaction clarified using model catalysts. *Science* **2016**, *351*, 361–365. [CrossRef]

21. Belova, A.I.; Kwabi, D.G.; Yashina, L.V.; Shao-Horn, Y.; Itkis, D.M. Mechanism of Oxygen Reduction in Aprotic Li–Air Batteries: The Role of Carbon Electrode Surface Structure. *J. Phys. Chem. C* **2017**, *121*, 1569–1577. [CrossRef]

22. Zhou, M.; Wang, H.-L.; Guo, S. Towards high-efficiency nanoelectrocatalysts for oxygen reduction through engineering advanced carbon nanomaterials. *Chem. Soc. Rev.* **2016**, *45*, 1273–1307. [CrossRef] [PubMed]

23. Wang, H.; Min, S.; Wang, Q.; Li, D.; Casillas, G.; Ma, C.; Li, Y.; Liu, Z.; Li, L.-J.; Yuan, J.; et al. Nitrogen-Doped Nanoporous Carbon Membranes with Co/CoP Janus-Type Nanocrystals as Hydrogen Evolution Electrode in Both Acidic and Alkaline Environments. *ACS Nano* **2017**, *11*, 4358–4364. [CrossRef] [PubMed]

24. Zagal, J.H.; Bedioui, F. *Electrochemistry of N4 Macrocyclic Metal Complexes: Volume 2: Biomimesis, Electroanalysis and Electrosynthesis of MN4 Metal Complexes*; Springer: Berlin, Germany, 2016.

25. Manbeck, G.F.; Fujita, E. A review of iron and cobalt porphyrins, phthalocyanines and related complexes for electrochemical and photochemical reduction of carbon dioxide. *J. Porphyr. Phthalocyanines* **2015**, *19*, 45–46. [CrossRef]

26. Wu, Z.-S.; Chen, L.; Liu, J.; Parvez, K.; Liang, H.; Shu, J.; Sachdev, H.; Graf, R.; Feng, X.; Müllen, K. High-Performance Electrocatalysts for Oxygen Reduction Derived from Cobalt Porphyrin-Based Conjugated Mesoporous Polymers. *Adv. Mater.* **2014**, *26*, 1450–1455. [CrossRef] [PubMed]

27. Chatterjee, S.; Sengupta, K.; Mondal, B.; Dey, S.; Dey, A. Factors Determining the Rate and Selectivity of $4e^-/4H^+$ Electrocatalytic Reduction of Dioxygen by Iron Porphyrin Complexes. *Acc. Chem. Res.* **2017**, *50*, 1744–1753. [CrossRef] [PubMed]

28. Hu, X.-M.; Salmi, Z.; Lillethorup, M.; Pedersen, E.B.; Robert, M.; Pedersen, S.U.; Skrydstrup, T.; Daasbjerg, K. Controlled electropolymerisation of a carbazole-functionalised iron porphyrin electrocatalyst for CO_2 reduction. *Chem. Commun.* **2016**, *52*, 5864–5867. [CrossRef]

29. Hod, I.; Farha, O.K.; Hupp, J.T. Electrocatalysis: Powered by porphyrin packing. *Nat. Mater.* **2015**, *14*, 1192–1193. [CrossRef]

30. Canales, C.; Ramírez, G. Glassy carbon electrodes modified with supramolecular assemblies generated by p-stacking of Cobalt (II) octaethylporphyrins. A 4 electrons-dioxygen reduction reaction occurring at positive potentials. *Electrochim. Acta* **2015**, *173*, 636–641. [CrossRef]

31. Canales, C.; Varas-Concha, F.; Mallouk, T.E.; Ramírez, G. Enhanced electrocatalytic hydrogen evolution reaction: Supramolecular assemblies of metalloporphyrins on glassy carbon electrodes. *Appl. Catal. B Environ.* **2016**, *188*, 169–176. [CrossRef]

32. Canales, C.; Olea, A.F.; Gidi, L.; Arce, R.; Ramírez, G. Enhanced light-induced hydrogen evolution reaction by supramolecular systems of cobalt(II) and copper(II) octaethylporphyrins on glassy carbon electrodes. *Electrochim. Acta* **2017**, *258*, 850–857. [CrossRef]

33. Tang, H.; Yin, H.; Wang, J.; Yang, N.; Wang, D.; Tan, Z. Molecular Architecture of Cobalt Porphyrin Multilayers on Reduced Graphene Oxide Sheets for High-Performance Oxygen Reduction Reaction. *Angew. Chem.* **2013**, *125*, 5695–5699. [CrossRef]

34. Jiang, R.; Chu, D. Comparative study of CoFeNx/C catalyst obtained by pyrolysis of hemin and cobalt porphyrin for catalytic oxygen reduction in alkaline and acidic electrolytes. *J. Power Sources* **2014**, *245*, 352–361. [CrossRef]

35. Oldacre, A.N.; Friedman, A.E.; Cook, T.R. A Self-Assembled Cofacial Cobalt Porphyrin Prism for Oxygen Reduction Catalysis. *J. Am. Chem. Soc.* **2017**, *139*, 1424–1427. [CrossRef] [PubMed]

36. Ma, T.Y.; Dai, S.; Jaroniec, M.; Qiao, S.Z. Metal−Organic Framework Derived Hybrid Co_3O_4-Carbon Porous Nanowire Arrays as Reversible Oxygen Evolution Electrodes. *J. Am. Chem. Soc.* **2014**, *136*, 13925–13931. [CrossRef] [PubMed]

37. Zhuang, Z.; Sheng, W.; Yan, Y. Synthesis of Monodispere $Au@Co_3O_4$ Core-Shell Nanocrystals and Their Enhanced Catalytic Activity for Oxygen Evolution Reaction. *Adv. Mater.* **2014**, *26*, 3950–3955. [CrossRef] [PubMed]

38. Scholz, F. (Ed.) *Electroanalytical Methods: Guide to Experiments and Applications*; Springer: Berlin, Germany, 2013; ISBN 3662047578, 9783662047576.

39. Marín, A.; Aguirre, M.J.; Muena, J.P.; Dehaen, W.; Maes, W.; Ngo, T.H.; Ramírez, G.; Arévalo, M.C. Electro-reduction of oxygen to water mediated by stable glassy carbon electrodes modified by Co(II)-porphyrins with voluminous meso-substituent. *Int. J. Electrochem. Sci.* **2015**, *10*, 3949.

40. Bard, A.J.; Faulkner, L.R. *Electrochemical Methods: Fundamentals and Applications*; Wiley: New York, NY, USA, 2000; ISBN 978-0-471-04372-0.

41. Ríos, R.; Marín, A.; Ramírez, G. Nitrite electro-oxidation mediated by Co(II)-[tetra(4-aminophenyl) porphyrin]-modified electrodes: Behavior as an amperometric sensor. *J. Coord. Chem.* **2010**, *63*, 1283–1294. [CrossRef]

42. Tham, M.K.; Walker, R.D.; Gubbins, K.E. Diffusion of Oxygen and Hydrogen in Aqueous Potassium Hydroxide Solutions. *J. Phys. Chem.* **1970**, *74*, 1747–1751. [CrossRef]

43. American Public Health Association (APHA); American Water Works Association (AWWA); Water Environment Federation (WEF). *Standard Methods for the Examination of Water and Wastewater*, 22nd ed.; United Book Press Inc.: Baltimore, MD, USA, 2012.

44. Andrieux, C.P.; Seavant, J.M. Heterogeneous versus homogeneous catalysis of electrochemical reactions. *J. Electroanal. Chem.* **1978**, *93*, 163–168. [CrossRef]

45. Srejic, I.; Rakocevic, Z.; Nenadovic, M.; Strbac, S. Oxygen reduction on polycrystalline palladium in acid and alkaline solutions: Topographical and chemical Pd surface changes. *Electrochim. Acta* **2015**, *169*, 22–31. [CrossRef]

46. Blizanac, B.; Ross, P.; Markovic, N. Oxygen Reduction on Silver Low-Index Single-Crystal Surfaces in Alkaline Solution: Rotating Ring DiskAg (h kl) Studies. *J. Phys. Chem. B* **2006**, *110*, 4735–4741. [CrossRef] [PubMed]

47. Stamenkovic, V.; Schmidt, T.; Ross, P.; Markovic, N. Surface composition effects in electrocatalysis: Kinetics of oxygen reduction on well-defined Pt_3Ni and Pt_3Co alloy surfaces. *J. Phys. Chem. B* **2002**, *106*, 11970–11979. [CrossRef]

48. Zhao, Y.; Liu, J.; Zhao, Y.; Wang, F. Composition-Controlled Synthesis of Carbon-Supported Pt–Co Alloy Nanoparticles and Their Origin of the Activity Enhancement for Oxygen Reduction Reaction. *Phys. Chem. Chem. Phys.* **2014**, *16*, 19298–19306. [CrossRef] [PubMed]

49. Córdoba-Torres, P.; Mesquita, T.J.; Nogueira, R.P. Relationship between the origin of constant-phase element behavior in electrochemical impedance spectroscopy and electrode surface structure. *J. Phys. Chem. C* **2015**, *119*, 4136–4147. [CrossRef]

50. Wang, S.; Zhang, L.; Xia, Z.; Roy, A.; Chang, D.W.; Baek, J.B.; Dai, L. BCN Graphene as Efficient Metal-Free Electrocatalyst for the Oxygen Reduction Reaction. *Angew. Chem.* **2012**, *124*, 4285–4288. [CrossRef]

51. Luo, Z.; Lim, S.; Tian, Z.; Shang, J.; Lai, L.; MacDonald, B.; Fu, C.; Shen, Z.; Yu, T.; Lin, J. Pyridinic N doped graphene: Synthesis, electronic structure, and electrocatalytic property. *J. Mater. Chem.* **2011**, *21*, 8038–8044. [CrossRef]

52. Li, Y.; Zhao, Y.; Cheng, H.; Hu, Y.; Shi, G.; Dai, L.; Qu, L. Nitrogen-Doped Graphene Quantum Dots with Oxygen-Rich Functional Groups. *J. Am. Chem. Soc.* **2012**, *134*, 15–18. [CrossRef]

53. Jiang, L.; Cui, L.; He, X. Cobalt-porphyrin noncovalently functionalized graphene as nonprecious-metal electrocatalyst for oxygen reduction reaction in an alkaline medium. *J. Solid State Electrochem.* **2015**, *19*, 497–506. [CrossRef]

54. Gidi, L.; Canales, C.; Aguirre, M.J.; Armijo, F.; Ramírez, G. Four-Electron Reduction of Oxygen Electrocatalyzed by a Mixture of Porphyrin Complexes onto Glassy Carbon Electrode. *Int. J. Electrochem. Sci.* **2018**, *13*, 1666–1682. [CrossRef]

 *catalysts*

Article

Heteroatom (Nitrogen/Sulfur)-Doped Graphene as an Efficient Electrocatalyst for Oxygen Reduction and Evolution Reactions

Jian Zhang [1], Jia Wang [2], Zexing Wu [2], Shuai Wang [2,*], Yumin Wu [1,*] and Xien Liu [2]

[1] College of Chemical Engineering, Qingdao University of Science & Technology, Qingdao 266042, China; jian8552@163.com

[2] State Key Laboratory Base of Eco-Chemical Engineering, College of Chemistry and Molecular Engineering, Qingdao University of Science & Technology, Qingdao 266042, China; wangjia3588@163.com (J.W.); splswzx@qust.edu.cn (Z.W.); liuxien@qust.edu.cn (X.L.)

* Correspondence: qustwangshuai@qust.edu.cn (S.W.); wuyumin001@126.com (Y.W.); Tel.: +86-150-9202-5911 (S.W.); +86-138-5463-8500 (Y.W.)

Received: 10 September 2018; Accepted: 18 October 2018; Published: 19 October 2018

Abstract: Carbon nanomaterials are potential materials with their intrinsic structure and property in energy conversion and storage. As the electrocatalysts, graphene is more remarkable in electrochemical reactions. Additionally, heteroatoms doping with metal-free materials can obtain unique structure and demonstrate excellent electrocatalytic performance. In this work, we proposed a facile method to prepare bifunctional electrocatalyst which was constructed by nitrogen, sulfur doped graphene (NSG), which demonstrate superior properties with high activity and excellent durability compared with Pt/C and IrO_2 for oxygen reduction (OR) and oxygen evolution (OE) reactions. Accordingly, these phenomena are closely related to the synergistic effect of doping with nitrogen and sulfur by rationally regulating the polarity of carbon in graphene. The current work expands the method towards carbon materials with heteroatom dopants for commercialization in energy-related reactions.

Keywords: electrocatalysts; bifunctional catalyst; graphene; dopants

---

1. Introduction

Graphene, a two-dimensional atom-thick conjugated structure, has drawn particular attention for its good conductivity, mechanical property, electrochemical stability and huge specific surface area as well as its wide potential applications in catalysts [1,2]. However, graphene oxide has serious structure disorder and contains oxygen-containing groups, which weaken its electrode transportation and conductivity [3–6]. Heteroatoms like N and S are adjacent elements to carbon. The doping of such elements to graphene may change its structure and improve the electrochemical property of graphene [4,7–10] due to its excellent stability, durability and controllability of nanoparticles [11,12].

On the other hand, fuel cells and metal-air batteries, as promising high-performance electrochemical energy-related devices, are suffering from bottlenecks because of its sluggish kinetics in oxygen reduction reaction (ORR, $O_2 + H_2O + 4e^- \rightarrow 4OH^-$) [13–18]. The platinum (Pt), with the advantage of low overpotential and high current density, has been served as the most promising ORR electrocatalyst regardless of the scarcity, high cost, and poor durability which obstruct its application in the energy field [7,19–22]. To address this problem, seeking alternative ORR and OER electrocatalysts with low overpotentials and cost is urgent for sustainable energy solutions. Heteroatom doped carbon-based materials as promising metal-free electrocatalysts have encouraged intensive research, while it is found that, the hybrids of doped graphene show multi-functions, and have a wide application prospect in energy storage and conversion and environmental detection. They can be used as catalysts for ORR and OER, but the relationship between precursor-doping pattern-ORR activity remains unclear [16,23–26]. Additionally, the performance of ORR and OER of some metal-free electrocatalysts can be optimized. Therefore, developing low cost and high electrocatalytic efficiency metal-free materials has become a focus of study.

Herein, we developed an efficient method to prepare heteroatom nitrogen-, sulfur-doped graphene as bifunctional materials for ORR/OER by hydrothermal synthesis of graphene oxide and potassium thiocyanate (KSCN), whether the guest gas (NH_3) exists or not.

2. Results and Discussion

2.1. Characterization of Electrocatalysts

The structure and morphology of materials were investigated through scanning electron microscopy (SEM) and transmission electron microscopy (TEM) [27]. As observed in Figure 1 and Figure S1, the elements of C, N, O and S were distributed uniformly in NSG, with the fabrication of stacking graphene sheets. This specific morphology could catalyze the reaction process derived from the exposing of more active sites. The SEM analyses (Figure 1b) confirm that the heteroatoms were doped successfully. The microstructure of nitrogen doped graphene (NG) and sulfur doped graphene (SG) were shown in Figures S2 and S3 where the twisted graphene layers were clearly observed.

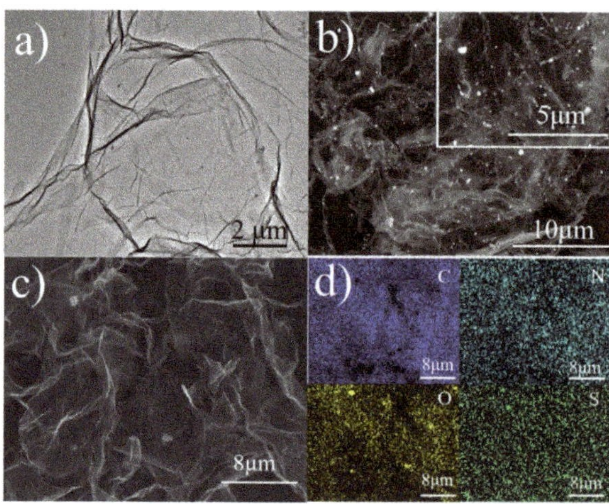

Figure 1. Transmission electron microscopy (TEM) (**a**), scanning electron microscopy (SEM) (**b**,**c**) and mapping (**d**) of C, N, O, S for NSG.

X-ray photoelectron spectroscopy (XPS) tests were the valid method to analyze the ingredients and chemical valence of materials [28]. As shown in Figure 2a, Compositions of C, N, O and S were evaluated through XPS survey spectrum whose atomic percentages respectively are 91.90, 4.18, 3.90 and 0.02 at %, demonstrating the successful doping with nitrogen and sulfur. As observed from Figure 2b, sp^2 hybridized graphitic carbon, C–S and C=O and C=N were corresponding to 284.5 eV, 285.5 eV and 287.0 eV, respectively. Pyridinic N (397.8 eV), pyrrolic N (398.9 eV) and graphitic N (400.3 eV) all existed (Figure 2c). As we know, graphitic N is viewed as active sites in the oxygen reduction reaction [23,28]. Moving to Figure 2d, high resolution S 2p, the peaks of 163.9 and 165.0 eV can ascribe to the S 2p spin-orbit doublet (S 2p1/2 and S 2p3/2) [29], affected by C–S–C, deeply demonstrating the successful addition with sulfur [30]. Compared with the high resolution N 1s of NG (Figure S4), the existence of C–S–C and synergistic effects of N and S doping may be the essential factors of performance improvement for NSG. All of this demonstrated the synergy of N and S catalyzed the activity of NSG.

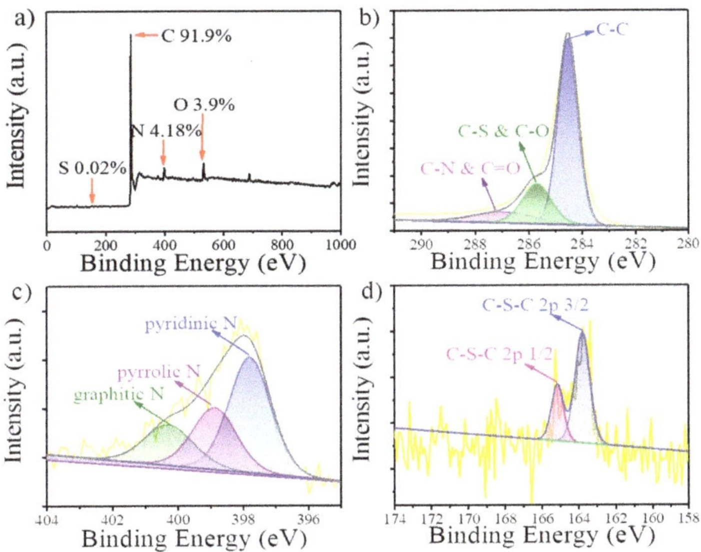

Figure 2. (**a**) Full range XPS spectra of NSG; (**b–d**) XPS spectrum of C 1s, N 1s and S 2p for NSG.

Figure 3a shows distinct Raman spectra at 1345 and 1600 cm^{-1} which is ascribed to the D and G band. A recent report indicated that the G band represents the graphitization degree of materials, structural defects are reflected by the D band [31]. To our best knowledge, the I_D/I_G ratio represents structural disorder in graphitic materials [32]. As observed from Figure 3a, the value of NSG (1.09) was lower than that of SG (1.15) and NG (1.29), indicating the increasing graphitic degree of NSG which may be derived from the "self-repairing" of intermediate products under experimental conditions with the repairing of partial sp^2-bonded C atoms. The components of the electrocatalysts were also analyzed. The graphitic C structure existed in NSG, SG and NG at 24° with the observation of X-ray diffraction (XRD) spectrum in Figure 3b [33]. In addition, for SG and NSG, the inconspicuous peak at 667 cm^{-1} belonged to the vibration of C–S bond which was making a clear indication that S element was doped with graphene in Figure S5 [5,8].

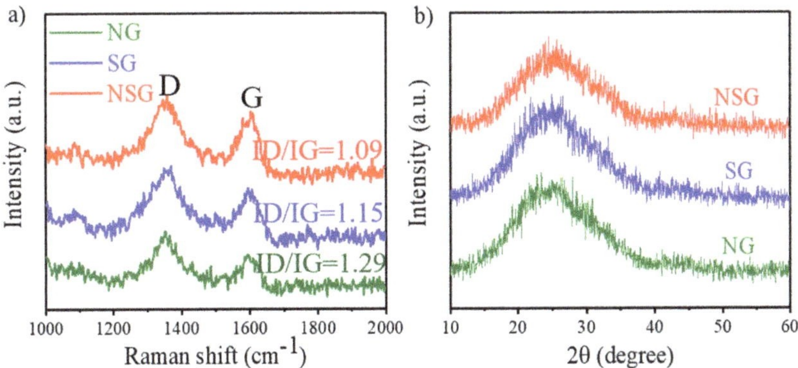

Figure 3. Raman spectra (**a**) and X-ray diffraction (XRD) (**b**) of NSG, NG and SG.

2.2. Electrocatalytic Properties of Catalysts

In O_2-saturated 0.1 M KOH electrolyte, a usual three-electrode electrochemical device consisted of Ag/AgCl, Pt wire and rotating ring-disk electrode (4 mm diameter) which was orderly used as a reference electrode, counter electrode and working electrode with the continuous flowing of O_2 operated to test the ORR performances of different catalysts which was revealed by linear sweep voltammetry (LSV) [34]. According to the conversion equation ($V_{RHE} = V_{Ag/AgCl} + 0.059pH + 0.197$), the potentials tested through Ag/AgCl were converted into reversible hydrogen electrode (RHE).

Compared with commercial Pt/C (20 wt %), the ORR performance of NSG, SG and NG was investigated via LSV in 0.1 M KOH electrolyte (Figure 4a). Obviously, NSG shows higher onset potential (0.95 V vs. RHE) compared with SG and NG (0.90 and 0.89 V). For half-wave potential, as shown in Figure 4b, the value of NSG (0.84 V) is prior to that of Pt/C, SG and NG (0.82, 0.8, 0.77 V), indicating its preponderant electrocatalytic performance for ORR. In addition, NSG displays large limited current density, demonstrating a more efficient mass transfer among such electrocatalysts [5].

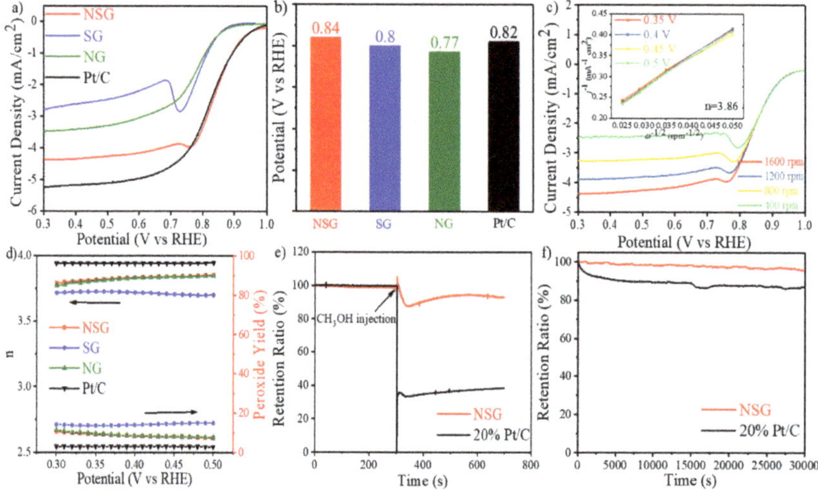

Figure 4. (**a**) Oxygen reduction reaction (ORR) polarization curves of Pt/C, NSG, SG, NG in O_2-saturated 0.1 M KOH solution, respectively (rotation speed 1600 rpm, sweep rate 10 mV s^{-1}); (**b**) Half-wave potential of NSG, SG, NG and Pt/C; (**c**) ORR polarization curves of NSG at the various rotation speeds (sweep rate 10 mV s^{-1}) (inset: Corresponding K-L plots at different electrode potentials); (**d**) The electron transfer number n of NSG, NG, SG and Pt/C at different potentials (left), and percentage (%) of peroxide with respect to the total oxygen reduction products (right); (**e**) Chronoamperometric response of NSG and 20% Pt/C at 0.57 V in O_2-saturated 0.1 mol L^{-1} KOH solution. The arrows indicate the addition of methanol; (**f**) Durability evaluation of NSG and 20% Pt/C at 0.57 V for 30,000 s with a rotating rate of 1600 rpm.

In order to quantitatively analyze the mechanism of synthesized catalysts for ORR, further LSVs were recorded at 400, 800, 1200 and 1600 rpm (Figure 4c, Figure S6a,c). Koutecky-Levich (K-L) plots were illustrated in Figure 4c and Figure S5b,d. To the best of our knowledge, a function relationship exists between inverse current density and square root of the rotation rate at a different range of potentials. According to the calculating formula of K-L equation, 3.86, 3.72 and 3.73, the electron transfer numbers (n) of NSG, SG and NG, indicate the ORRs mechanism is a four-electron process.

To further explore the performance of synthesized electrocatalysts, the electron transfer number (n) and percentage (%) of peroxide species were tested through a rotating ring disk electrode (RRDE). As shown in Figure 4d, the four-electron process of the as-prepared NSG for ORR is consistent with the

RDE results. Additionally the average percentage (%) of peroxide yield of NSG is the lowest compared with SG and NG, demonstrating the multiple doping.

Adding 8.5 mL methanol into 70 mL 0.1 M KOH electrolyte with continuous O_2, the tolerance of NSG to methanol was investigated. Compared with the sharp decline of current density of Pt/C (current loss of ~60%), NSG exhibits a little fluctuation with a retention ratio of ~93%, demonstrating the decreased activity of Pt/C affected by methanol, and NSG possesses a well tolerance to methanol (Figure 4e).

As a state-of-the-art catalyst for ORR, chronoamperometric measurement was used to assess the durability at a constant cathodic voltage of 0.57 V. As observed from Figure 4f, NSG exhibits an outstanding ORR stability with a weak attenuation over 30,000 s, maintaining 96% of the initial current. Compared with approximately 15% loss of initial current, NSG was better than Pt/C for ORR in alkaline electrolyte. OER activity of various as-prepared samples was also tested in the same condition (Figure 5a). To our best knowledge, the quantitative value of $E_j = 10$ (potentials to deliver 10 mA cm^{-2} current density) represents 10% efficient solar water-splitting cell which is used to make comparison with various catalysts [35]. For NSG, $E_j = 10$ (1.62 V) is lower than that of 1.65 V (SG), 1.66 V (NG) and some of state-of-the-art carbon-based catalysts, just like N, S-CN [5] and CN nanocables [36] and so on. The Tafel slope of NSG (105 mV dec^{-1}) is lower than that of SG (121 mV dec^{-1}) and NG (124 mV dec^{-1}), indicating its favorable reaction kinetics (Figure 5b). In addition, the representation of charge transfer kinetics, electrochemical impedance spectroscopy (EIS) of NSG, NG and SG, were also investigated. The Nyquist plots in Figure S7 demonstrates a lower charge transfer resistance towards the OER process of NSG than that of NG and SG, which is consistent with the OER performance, revealing faster Faradaic process in OER kinetics of NSG.

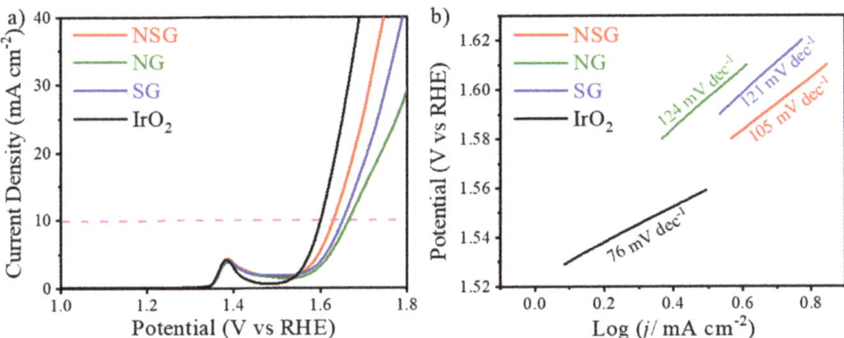

Figure 5. (**a**) OER linear sweeping voltammetrys (LSVs) of NG, SG, IrO$_2$ and NSG at a sweep rate of 10 mV s^{-1}; (**b**) OER Tafel plots.

Considering the structure-property relationship, the better catalytic performance towards ORR and OER may be a result of the stable covalent C–N which could form high positive charge density on neighboring carbon atoms, and the mismatch of outermost orbitals between C and S [37] and large surface area [38] which could facilitate the charge transfer, further endowing more accessible catalytic surfaces.

3. Materials and Methods

3.1. Preparation of Electrocatalysts

At first, GO was prepared according to the procedure used by Hummer [39]. To begin this process, GO and KSCN (20.0 mg/20.0 mg) were intensively mixed in 30.0 mL deionized water. Then, a homogeneous solution was fabricated through ultrasonication and kept in Teflon-lined stainless-steal autoclave (150 °C) for 15 h. The as-prepared solution was treated, at freezing, with a vacuum dryer overnight to form powder. After dried, the powder was placed in a tube furnace which was programmed at 800 °C for 2 h with 5 °C/min of rising rate, keeping Ar flowing of 150 mL/min. In the end, with Ar and NH_3 whose flow rate is 500 mL/min, the intermediate product was annealed at 800 °C for 1 h (Scheme 1). Up to now, the target materials were obtained and defined as nitrogen-, sulfur- co-doped graphene (NSG). For the sake of contrast, only S-doped or N-doped graphene catalysts were synthesized under the same operating conditions which was defined as SG or NG, respectively. In other words, the only source of N (NH_3) or S (KSCN) was used as introduction.

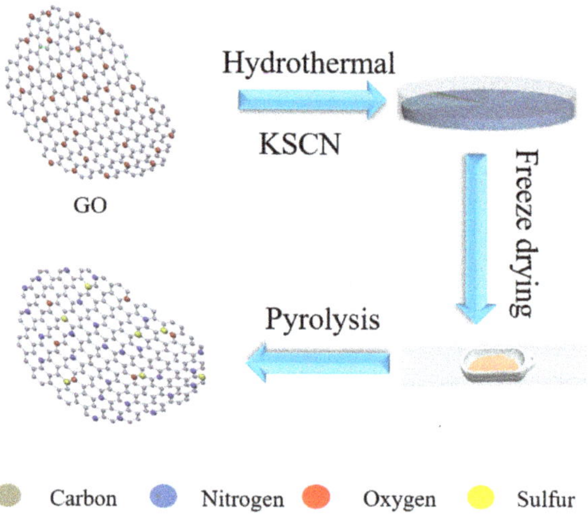

Scheme 1. Schematic illustration of the preparation of NSG.

3.2. Characterization of Electrocatalysts

A S-4800 SEM instrument (Hitachi High-Technology Co., Ltd., Tokyo, Japan) was used to test the surface characterization of different electrocatalysts. TEM (JEOL JEM-2100, Tokyo, Japan) was operated at 200 kV. With radiation of Cu-Kα, X-ray diffraction (XRD, D/Max2000, Rigaku, Japan) was investigated. Fourier transformed infrared (FTIR) spectra were observed through a TENSOR 27 FT-IR spectrometer (Scotia, NY, USA) in the range of 4000–500 cm^{-1}, after the samples were dried. Escalab 250 xi (Thermo Scientific, Loughborough, UK) was used to record X-ray photoelectron spectroscopy (XPS), providing a base pressure of 5×10^{-9} Torr radiated from monochromatic Al Kα. Raman spectra were investigated by using a Renishaw Raman spectroscope (Renishaw plc., Gloucestershire, UK).

4. Conclusions

To conclude this work, the four-electron pathway for ORR on N, S-co-doped graphene is revealed and synergy effects between dopants are discussed. The synergy effect is ascribed to the increasing spin density with the dopant of S and graphitic N. The as-prepared catalyst exhibits excellent performance for ORR and OER which is originated from the unique structure of NSG, fortunately, the unique structure is to the benefit of mass transfer. Overall, this work provides a carbon-based bifunctional electrocatalyst of dual doping in ORR and OER on the promising widespread application in energy-related devices.

Supplementary Materials: The following are available online at http://www.mdpi.com/2073-4344/9/10/475/s1, Figure S1: Energy dispersive spectrometer (EDS) of NSG, Figure S2: SEM (a-c) and mapping of C (d), N (e), O (f) of NG, Figure S3: SEM (a,b) and mapping of C (c), N (d), O (e) and S (f) of SG, Figure S4: Full range XPS spectra of NG, XPS spectrum of C 1s and N 1s for NG, Figures S5: Fourier transform infrared spectroscopy (FTIR) of NSG,

SG, NG and GO, Figure S6: (a,c) Linear Scan Voltammetry (LSV) curves for SG and NG at different rotation rates in 0.1 M KOH. (b,d) Crresponding K-L plots at different potentials: 0.35, 0.4, 0.45, 0.5 V, Figure S7: Nyquist plots of electrochemical impedance spectra (EIS) of NSG, SG and NG recorded in 1 M KOH. Inset: One-time-constant model equivalent circuit used for data fitting of EIS spectra, Table S1: Comparison of ORR and OER performance of NSG with the recently reported metal-free catalysts at 1600 rpm in KOH solution.

Author Contributions: Data curation, J.Z.; Project administration, X.L.; Software, J.W.; Supervision, Z.W.; Writing—original draft, S.W.; Writing—review and editing, Y.W.

Funding: This research was funded by Taishan Scholar Program of Shandong Province, China (ts201712045). The Key Research and Development Program of Shandong Province (2018GGX104001). Natural Science Foundation of Shandong Province of China (ZR2017MB054). Doctoral Fund of QUST (0100229001). Post-doctoral Applied Research Fund of Qingdao (04000641). (∓, equally contribution).

Conflicts of Interest: The authors declare no conflict of interest.

References

1. Higgins, D.; Hoque, M.A.; Seo, M.H.; Wang, R.; Hassan, F.; Choi, J.-Y.; Pritzker, M.; Yu, A.; Zhang, J.; Chen, Z. Development and simulation of sulfur-doped graphene supported platinum with exemplary stability and activity towards oxygen reduction. *Adv. Funct. Mater.* **2014**, *24*, 4325–4336. [CrossRef]
2. Ge, X.; Sumboja, A.; Wuu, D.; An, T.; Li, B.; Goh, F.W.T.; Hor, T.S.A.; Zong, Y.; Liu, Z. Oxygen reduction in alkaline media: From mechanisms to recent advances of catalysts. *ACS Catal.* **2015**, *5*, 4643–4667. [CrossRef]
3. Han, J.H.; Huang, G.; Wang, Z.L.; Lu, Z.; Du, J.; Kashani, H.; Chen, M.W. Low temperature carbide-mediated growth of bicontinuous nitrogen-doped mesoporous graphene as an efficient oxygen reduction electrocatalyst. *Adv. Mater.* **2018**, *30*, 1803588. [CrossRef] [PubMed]
4. Li, J.; Zhang, Y.; Zhang, X.; Huang, J.; Han, J.; Zhang, Z.; Han, X.; Xu, P.; Song, B.S. N dual-doped graphene-like carbon nanosheets as efficient oxygen reduction reaction electrocatalysts. *ACS Appl. Mater. Interfaces* **2017**, *9*, 398–405. [CrossRef] [PubMed]
5. Qu, K.; Zheng, Y.; Dai, S.; Qiao, S.Z. Graphene oxide-polydopamine derived n, s-codoped carbon nanosheets as superior bifunctional electrocatalysts for oxygen reduction and evolution. *Nano Energy* **2016**, *19*, 373–381. [CrossRef]
6. Sakthinathan, S.; Kubendhiran, S.; Chen, S.-M.; Karuppiah, C.; Chiu, T.-W. Novel bifunctional electrocatalyst for orr activity and methyl parathion detection based on reduced graphene oxide/palladium tetraphenylporphyrin nanocomposite. *J. Phys. Chem. C* **2017**, *121*, 14096–14107. [CrossRef]
7. Wu, J.; Ma, L.; Yadav, R.M.; Yang, Y.; Zhang, X.; Vajtai, R.; Lou, J.; Ajayan, P.M. Nitrogen-doped graphene with pyridinic dominance as a highly active and stable electrocatalyst for oxygen reduction. *ACS Appl. Mater. Interfaces* **2015**, *7*, 14763–14769. [CrossRef] [PubMed]
8. Akhter, T.; Islam, M.M.; Faisal, S.N.; Haque, E.; Minett, A.I.; Liu, H.K.; Konstantinov, K.; Dou, S.X. Self-assembled n/s codoped flexible graphene paper for high performance energy storage and oxygen reduction reaction. *ACS Appl. Mater. Interfaces* **2016**, *8*, 2078–2087. [CrossRef] [PubMed]
9. Men, B.; Sun, Y.; Liu, J.; Tang, Y.; Chen, Y.; Wan, P.; Pan, J. Synergistically enhanced electrocatalytic activity of sandwich-like n-doped graphene/carbon nanosheets decorated by fe and s for oxygen reduction reaction. *ACS Appl. Mater. Interfaces* **2016**, *8*, 19533–19541. [CrossRef] [PubMed]
10. Khilari, S.; Pradhan, D. Mnfe2o4@nitrogen-doped reduced graphene oxide nanohybrid: An efficient bifunctional electrocatalyst for anodic hydrazine oxidation and cathodic oxygen reduction. *Catal. Sci. Technol.* **2017**, *7*, 5920–5931. [CrossRef]
11. Alshehri, S.M.; Alhabarah, A.N.; Ahmed, J.; Naushad, M.; Ahamad, T. An efficient and cost-effective tri-functional electrocatalyst based on cobalt ferrite embedded nitrogen doped carbon. *J. Colloid Interface Sci.* **2018**, *514*, 1–9. [CrossRef] [PubMed]
12. Alshehri, S.M.; Ahmed, J.; Ahamad, T.; Alhokbany, N.; Arunachalam, P.; Almayouf, A.M.; Ahmad, T. Synthesis, characterization, multifunctional electrochemical (OGR/ORR/SCs) and photodegradable activities of ZnWO$_4$ nanobricks. *J. Sol-Gel Sci. Technol.* **2018**, *87*, 137–146. [CrossRef]
13. Hoque, M.A.; Hassan, F.M.; Higgins, D.; Choi, J.Y.; Pritzker, M.; Knights, S.; Ye, S.; Chen, Z. Multigrain platinum nanowires consisting of oriented nanoparticles anchored on sulfur-doped graphene as a highly active and durable oxygen reduction electrocatalyst. *Adv. Mater.* **2015**, *27*, 1229–1234. [CrossRef] [PubMed]

14. Li, Y.; Gong, M.; Liang, Y.; Feng, J.; Kim, J.E.; Wang, H.; Hong, G.; Zhang, B.; Dai, H. Advanced zinc-air batteries based on high-performance hybrid electrocatalysts. *Nat. Commun.* **2013**, *4*, 1805. [CrossRef] [PubMed]

15. Lv, Y.; Wang, X.; Mei, T.; Li, J.; Wang, J. Single-step hydrothermal synthesis of n, s-dual-doped graphene networks as metal-free efficient electrocatalysts for oxygen reduction reaction. *ChemistrySelect* **2018**, *3*, 3241–3250. [CrossRef]

16. Gaidukevič, J.; Razumienė, J.; Šakinytė, I.; Rebelo, S.L.H.; Barkauskas, J. Study on the structure and electrocatalytic activity of graphene-based nanocomposite materials containing $(SCN)_n$. *Carbon* **2017**, *118*, 156–167. [CrossRef]

17. Song, J.; Liu, T.; Ali, S.; Li, B.; Su, D. The synergy effect and reaction pathway in the oxygen reduction reaction on the sulfur and nitrogen dual doped graphene catalyst. *Chem. Phys. Lett.* **2017**, *677*, 65–69. [CrossRef]

18. Alshehri, S.M.; Ahmed, J.; Ahamad, T.; Arunachalam, P.; Ahmad, T.; Khan, A. Bifunctional electro-catalytic performances of $CoWO_4$ nanocubes for water redox reactions (OER/ORR). *RSC Adv.* **2017**, *7*, 45615–45623. [CrossRef]

19. Ghanem, M.A.; Arunachalam, P.; Almayouf, A.; Weller, M.T. Efficient Bi-Functional Electrocatalysts of Strontium Iron Oxy-Halides for Oxygen Evolution and Reduction Reactions in Alkaline Media. *J. Electrochem. Soc.* **2016**, *163*, H450–H458. [CrossRef]

20. Chang, Y.; Hong, F.; He, C.; Zhang, Q.; Liu, J. Nitrogen and sulfur dual-doped non-noble catalyst using fluidic acrylonitrile telomer as precursor for efficient oxygen reduction. *Adv. Mater.* **2013**, *25*, 4794–4799. [CrossRef] [PubMed]

21. Wang, L.; Ambrosi, A.; Pumera, M. "Metal-free" catalytic oxygen reduction reaction on heteroatom- doped graphene is caused by trace metal impurities. *Angew. Chem. Int. Ed.* **2013**, *52*, 13818–13821. [CrossRef] [PubMed]

22. Wang, N.; Lu, B.; Li, L.; Niu, W.; Tang, Z.; Kang, X.; Chen, S. Graphitic nitrogen is responsible for oxygen electroreduction on nitrogen-doped carbons in alkaline electrolytes: Insights from activity attenuation studies and theoretical calculations. *ACS Catal.* **2018**, *8*, 6827–6836. [CrossRef]

23. Chai, G.-L.; Qiu, K.; Qiao, M.; Titirici, M.-M.; Shang, C.; Guo, Z. Active sites engineering leads to exceptional orr and oer bifunctionality in p,n co-doped graphene frameworks. *Energy Environ. Sci.* **2017**, *10*, 1186–1195. [CrossRef]

24. Chenitz, R.; Kramm, U.I.; Lefèvre, M.; Glibin, V.; Zhang, G.; Sun, S.; Dodelet, J.-P. A specific demetalation of $Fe-N_4$ catalytic sites in the micropores of NC_Ar + NH_3 is at the origin of the initial activity loss of the highly active Fe/N/C catalyst used for the reduction of oxygen in pem fuel cells. *Energy Environ. Sci.* **2018**, *11*, 365–382. [CrossRef]

25. Hassani, S.S.; Samiee, L.; Ghasemy, E.; Rashidi, A.; Ganjali, M.R.; Tasharrofi, S. Porous nitrogen-doped graphene prepared through pyrolysis of ammonium acetate as an efficient orr nanocatalyst. *Int. J. Hydrogen Energy* **2018**, *43*, 15941–15951. [CrossRef]

26. Kumar, K.; Canaff, C.; Rousseau, J.; Arrii-Clacens, S.; Napporn, T.W.; Habrioux, A.; Kokoh, K.B. Effect of the oxide–carbon heterointerface on the activity of Co_3O_4/nrgo nanocomposites toward ORR and OER. *J. Phys. Chem. C* **2016**, *120*, 7949–7958. [CrossRef]

27. Higgins, D.; Zamani, P.; Yu, A.; Chen, Z. The application of graphene and its composites in oxygen reduction electrocatalysis: A perspective and review of recent progress. *Energy Environ. Sci.* **2016**, *9*, 357–390. [CrossRef]

28. Yan, P.; Liu, J.; Yuan, S.; Liu, Y.; Cen, W.; Chen, Y. The promotion effects of graphitic and pyridinic n combinational doping on graphene for orr. *Appl. Surf. Sci.* **2018**, *445*, 398–403. [CrossRef]

29. Li, F.; Sun, L.; Luo, Y.; Li, M.; Xu, Y.; Hu, G.; Li, X.; Wang, L. Effect of thiophene s on the enhanced orr electrocatalytic performance of sulfur-doped graphene quantum dot/reduced graphene oxide nanocomposites. *RSC Adv.* **2018**, *8*, 19635–19641. [CrossRef]

30. Hoque, M.A.; Hassan, F.M.; Jauhar, A.M.; Jiang, G.; Pritzker, M.; Choi, J.-Y.; Knights, S.; Ye, S.; Chen, Z. Web-like 3d architecture of pt nanowires and sulfur-doped carbon nanotube with superior electrocatalytic performance. *ACS Sustain. Chem. Eng.* **2017**, *6*, 93–98. [CrossRef]

31. Chen, S.; Duan, J.; Jaroniec, M.; Qiao, S.Z. Nitrogen and oxygen dual-doped carbon hydrogel film as a substrate-free electrode for highly efficient oxygen evolution reaction. *Adv. Mater.* **2014**, *26*, 2925–2930. [CrossRef] [PubMed]

32. Lei, W.; Liu, S.; Zhang, W.-H. Porous hollow carbon nanofibers derived from multi-walled carbon nanotubes and sucrose as anode materials for lithium-ion batteries. *RSC Adv.* **2017**, *7*, 224–230. [CrossRef]

33. Cazetta, A.L.; Zhang, T.; Silva, T.L.; Almeida, V.C.; Asefa, T. Bone char-derived metal-free n- and s-co-doped nanoporous carbon and its efficient electrocatalytic activity for hydrazine oxidation. *Appl. Catal. B Environ.* **2018**, *225*, 30–39. [CrossRef]

34. Xu, C.; Xu, B.; Gu, Y.; Xiong, Z.; Sun, J.; Zhao, X.S. Graphene-based electrodes for electrochemical energy storage. *Energy Environ. Sci.* **2013**, *6*, 1388. [CrossRef]

35. Paulraj, A.; Kiros, Y.; Göthelid, M.; Johansson, M. Nifeox as a bifunctional electrocatalyst for oxygen reduction (OR) and evolution (OE) reaction in alkaline media. *Catalysts* **2018**, *8*, 328. [CrossRef]

36. Tian, G.-L.; Zhang, Q.; Zhang, B.; Jin, Y.-G.; Huang, J.-Q.; Su, D.S.; Wei, F. Toward full exposure of "active sites": Nanocarbon electrocatalyst with surface enriched nitrogen for superior oxygen reduction and evolution reactivity. *Adv. Funct. Mater.* **2014**, *24*, 5956–5961. [CrossRef]

37. Huang, H.; Ma, L.; Tiwary, C.S.; Jiang, Q.; Yin, K.; Zhou, W.; Ajayan, P.M. Worm-shape pt nanocrystals grown on nitrogen-doped low-defect graphene sheets: Highly efficient electrocatalysts for methanol oxidation reaction. *Small* **2017**, *13*, 1603013. [CrossRef]

38. Qiao, X.; Jin, J.; Fan, H.; Cui, L.; Ji, S.; Li, Y.; Liao, S. Cobalt and nitrogen co-doped graphene-carbon nanotube aerogel as an efficient bifunctional electrocatalyst for oxygen reduction and evolution reactions. *Catalysts* **2018**, *8*, 275. [CrossRef]

39. Yang, Q.; Su, Y.; Chi, C.; Cherian, C.T.; Huang, K.; Kravets, V.G.; Wang, F.C.; Zhang, J.C.; Pratt, A.; Grigorenko, A.N.; et al. Ultrathin graphene-based membrane with precise molecular sieving and ultrafast solvent permeation. *Nat. Mater.* **2017**, *16*, 1198–1202. [CrossRef] [PubMed]

MDPI
St. Alban-Anlage 66
4052 Basel
Switzerland
Tel. +41 61 683 77 34
Fax +41 61 302 89 18
www.mdpi.com

Catalysts Editorial Office
E-mail: catalysts@mdpi.com
www.mdpi.com/journal/catalysts